Water Science and Technology Library

Volume 99

The aim of the *Water Science and Technology Library* is to provide a forum for dissemination of the state-of-the-art of topics of current interest in the area of water science and technology. This is accomplished through publication of reference books and monographs, authored or edited. Occasionally also proceedings volumes are accepted for publication in the series. *Water Science and Technology Library* encompasses a wide range of topics dealing with science as well as socio-economic aspects of water, environment, and ecology. Both the water quantity and quality issues are relevant and are embraced by *Water Science and Technology Library*. The emphasis may be on either the scientific content, or techniques of solution, or both. There is increasing emphasis these days on processes and *Water Science and Technology Library* is committed to promoting this emphasis by publishing books emphasizing scientific discussions of physical, chemical, and/or biological aspects of water resources. Likewise, current or emerging solution techniques receive high priority. Interdisciplinary coverage is encouraged. Case studies contributing to our knowledge of water science and technology are also embraced by the series. Innovative ideas and novel techniques are of particular interest.

Comments or suggestions for future volumes are welcomed.

Vijay P. Singh, Department of Biological and Agricultural Engineering & Zachry Department of Civil and Environment Engineering, Texas A&M University, USA
Email: vsingh@tamu.edu

More information about this series at http://www.springer.com/series/6689

Taesam Lee · Vijay P. Singh ·
Kyung Hwa Cho

Deep Learning for Hydrometeorology and Environmental Science

Springer

Taesam Lee
Department of Civil Engineering
Gyeongsang National University
Jinju, Korea (Republic of)

Kyung Hwa Cho
School of Urban and Environmental Engineering
Ulsan National Institute of Science and Technology
Ulsan, Korea (Republic of)

Vijay P. Singh
Department of Biological and Agricultural Engineering, Zachry Department of Civil and Environmental Engineering
Texas A&M University
College Station, TX, USA

ISSN 0921-092X ISSN 1872-4663 (electronic)
Water Science and Technology Library
ISBN 978-3-030-64776-6 ISBN 978-3-030-64777-3 (eBook)
https://doi.org/10.1007/978-3-030-64777-3

This Springer imprint is published by the registered company Springer Nature Switzerland AG
The registered company address is: Gewerbestrasse 11, 6330 Cham, Switzerland

Dedicated to

TL: the memory of my mom, Gumran Cho

VPS: my wife Anita, who is no more, son Vinay, daughter Arti, daughter-in-law Sonali, son-in-law Vamsi, and grandchildren Ronin, Kayden, and Davin

KC: Wife Yeonju and Daughter Yuna

Preface

Deep learning is known as part of a machine learning methodology based on an artificial neural network. Increasing data availability and computing power enhance applications of deep learning to hydrometeorological and environmental fields. However, books that specifically focus on the application to these fields are limited. Therefore, this book focuses on the explanation of deep learning techniques and their applications to hydrometeorological and environmental studies.

This book is divided into three parts. The first part is the introduction of the basic neural network, covering the basic concepts of artificial neural network in Chaps. 1–7. Chapter 1 introduces the concept of deep learning, followed by the mathematical background in Chap. 2. In Chap. 3, how to preprocess a dataset before applying a model is presented. Chapter 4 describes the terminology and structure of neural network models. The procedure of training a neural network is discussed in Chap. 5. The approaches to update the weights of a network model are presented in Chap. 6. The techniques to improve the model performance are given in Chap. 7.

The second part introduces advanced techniques in deep learning algorithms from Chaps. 8–10. The advanced neural network algorithms, as Extreme Learning Machine and Autoencoding, are presented in Chap. 8. The temporal deep learning techniques, as Long Short-term Memory (LSTM) and Gated Recurrent Unit (GRU), are discussed in Chap. 9. The spatial deep learning technique, as Convolution Neural Network (CNN), is introduced in Chap. 10.

The third part illustrates how to apply deep learning techniques to real case studies. In Chap. 11, Tensor flow and Keras programming is presented to illustrate how to simply implement deep learning to real datasets. Hydrometeorological and environmental applications of deep learning models are presented in Chaps. 12 and 13, respectively.

The book will be useful to graduate students, college faculty, and researchers in hydrology, meteorology, and environmental sciences. It may also be useful to policymakers in government at local, state, and national levels.

The first author acknowledges his former student, Mrs. Mahsa Moradi, for providing excellent hydrometeorological deep learning examples and the National Research Foundation of Korea for providing partial fund for the current work.

Jinju, South Korea — Taesam Lee
College Station, TX, USA — Vijay P. Singh
Ulsan, South Korea — Kyung Hwa Cho
June 2020

Contents

1 Introduction ... 1
1.1 What is Deep Learning? ... 1
1.2 Pros and Cons of Deep Learning ... 2
1.3 Recent Applications of Deep Learning in Hydrometeorological and Environmental Studies ... 2
1.4 Organization of Chapters ... 3
1.5 Summary and Conclusion ... 4
References ... 4

2 Mathematical Background ... 5
2.1 Linear Regression Model ... 5
2.1.1 Simple Linear Regression ... 5
2.1.2 Multiple Linear Regression ... 7
2.2 Time Series Model ... 10
2.2.1 Autoregressive Model (AR) ... 10
2.3 Probability Distributions ... 14
2.3.1 Normal Distributions ... 14
2.3.2 Gamma Distribution ... 17
2.4 Exercises ... 18
References ... 19

3 Data Preprocessing ... 21
3.1 Normalization ... 21
3.2 Data Splitting for Training and Testing ... 24
3.3 Exercises ... 25

4 Neural Network ... 27
4.1 Terminology in Neural Network ... 27
4.1.1 Components of Neural Network ... 27
4.1.2 Activation Functions ... 28

4.1.3 Error and Loss Function 33
4.1.4 Softmax and One-Hot Encoding 36
4.2 Artificial Neural Network 40
4.2.1 Simplest Network 40
4.2.2 Feedforward and Backward Propagation 41
4.2.3 Network with Multiple Input and Output Variables . . . 42
4.2.4 Python Coding of the Simple Network 44
4.3 Exercises 46

5 Training a Neural Network 47
5.1 Initialization 47
5.2 Gradient Descent 49
5.3 Backpropagation 51
5.3.1 Simple Network 51
5.3.2 Full Neural Network 54
5.3.3 Python Coding of Network 60
5.4 Exercises 62
Reference 62

6 Updating Weights 63
6.1 Momentum 63
6.2 Adagrad 65
6.3 RMSprop 68
6.4 Adam 70
6.5 Nadam 74
6.6 Python Coding of Updating Weights 76
6.7 Exercises 78
References 78

7 Improving Model Performance 79
7.1 Batching and Minibatch 79
7.2 Validation 80
7.2.1 Python Coding of K-Fold Cross-Validation 80
7.3 Regularization 81
7.3.1 L-Norm Regularization 81
7.3.2 Dropout 82
7.3.3 Python Coding of Regularization 84
7.4 Exercises 85
Reference 86

8 Advanced Neural Network Algorithms 87
8.1 Extreme Learning Machine (ELM) 87
8.1.1 Basic ELM 89
8.1.2 Generalized ELM 90
8.1.3 Python Coding 94

8.2 Autoencoder . . . 95
8.2.1 Vanilla Autoencoder . . . 97
8.2.2 Regularized Autoencoder . . . 102
8.2.3 Python Coding of Regularized AE . . . 104
8.3 Exercises . . . 106
Reference . . . 106

9 Deep Learning for Time Series . . . 107
9.1 Recurrent Neural Network . . . 107
9.1.1 Backpropagation . . . 108
9.1.2 Backpropagation Through Time (BPTT) . . . 109
9.2 Long Short-Term Memory (LSTM) . . . 113
9.2.1 Basics of LSTM . . . 113
9.2.2 Example of LSTM . . . 115
9.2.3 Backpropagation of a Simple LSTM . . . 117
9.2.4 Backpropagation Through Time (BPTT) . . . 120
9.3 Gated Recurrent Unit (GRU) . . . 124
9.3.1 Basics of GRU . . . 124
9.3.2 Example of GRU . . . 125
9.3.3 Backpropagation of a Simple GRU Model . . . 127
9.4 Exercises . . . 131
References . . . 131

10 Deep Learning for Spatial Datasets . . . 133
10.1 Convolutional Neural Network (CNN) . . . 133
10.1.1 Definition of Convolution . . . 133
10.1.2 Elements of CNN . . . 136
10.2 Backpropagation of CNN . . . 139
10.3 Exercises . . . 149

11 Tensorflow and Keras Programming for Deep Learning . . . 151
11.1 Basic Keras Modeling . . . 151
11.2 Temporal Deep Learning (LSTM and GRU) . . . 154
11.3 Spatial Deep Learning (CNN) . . . 159
11.4 Exercises . . . 162
References . . . 162

12 Hydrometeorological Applications of Deep Learning . . . 163
12.1 Stochastic Simulation with LSTM . . . 163
12.1.1 Mathematical Description for Stochastic Simulation with LSTM . . . 163
12.1.2 Colorado Monthly Streamflow . . . 164
12.1.3 Results of Colorado River . . . 164
12.1.4 Python Coding . . . 167
12.1.5 Matlab Coding . . . 170

12.2 Forecasting Daily Temperature with LSTM 174
12.2.1 Preparing the Data 175
12.2.2 Methodology 175
12.2.3 Results 177
12.2.4 Python Coding 178
12.3 Exercises 190
References 190

13 Environmental Applications of Deep Learning 191
13.1 Remote Sensing of Water Quality Using CNN 191
13.1.1 Introduction 191
13.1.2 Study Area and Monitoring 192
13.1.3 Field Data Collection 193
13.1.4 Point-Centered Regression CNN (PRCNN) 195
13.1.5 Results and Discussion 197
13.1.6 Conclusion 199
13.1.7 Python Coding 200
References 203

About the Authors

Prof. Taesam Lee, Ph.D. is a Full Professor in the Department of Civil Engineering at Gyeongsang National University in Jinju, South Korea. He got his Ph.D. degree from Colorado State University with a stochastic simulation of stream flow. He specializes in surface-water hydrology, meteorology, machine learning algorithms, and climatic changes in hydrological extremes publishing around 50 technical papers and a statistical downscaling book. He is a member of the American Society of Civil Engineers (ASCE) and the American Geophysical Union (AGU) and the associate editor of the *Journal of Hydrologic Engineering* in ASCE.

Prof. Vijay P. Singh is a University Distinguished Professor, a Regents Professor, and Caroline and William N. Lehrer Distinguished Chair in Water Engineering at Texas A&M University. He received his B.S., M.S., Ph.D. and D.Sc. degrees in engineering. He is a registered professional engineer, a registered professional hydrologist, and an Honorary Diplomate of ASCE-AAWRE. He has published more than 1270 journal articles, 30 textbooks, 70 edited reference books, 105 book chapters, and 315 conference papers in the area of hydrology and water resources. He has received more than 90 national and international awards, including three honorary doctorates. He is a member of 11 international science/engineering academies. He has served as President of the American Institute of Hydrology (AIH), Chair of Watershed Council of American Society of Civil Engineers, and is currently President-Elect of the American Academy of Water Resources Engineers. He has served/serves as editor-in-chief of three journals and two book series and serves on editorial boards of more than 25 journals and three book series.

Prof. Kyung Hwa Cho, Ph.D. is an Associate Professor in the School of Urban and Environmental Engineering at Ulsan National Institute of Science and Technology, South Korea. He obtained his B.S. in chemical engineering and M.S. and Ph.D. in Environmental Engineering. He has published more than 110 journal articles in water and environmental journals such as *Water Research*, *Remote Sensing of Environment*. His expertise lies in modeling water quality, deep learning application for water quality prediction, and using hyperspectral images for water quality monitoring.

Chapter 1
Introduction

Abstract Deep learning has been popularly employed for analysis and forecasting in various fields. In this chapter, a brief introduction of deep learning is presented, including the definition and pros and cons of deep learning, followed by the recent applications of deep learning models in hydrological and environmental fields. The structure of the remaining chapters for this book is also explained.

1.1 What is Deep Learning?

In recent years, deep learning techniques have been developed and employed in a number of fields, such as voice search, automatic text generation, health care, and image recognition. The skyrocketing development and applications of deep learning stem from its capability of image classification and object detection with much more accuracy than a human can do.

Old-fashioned machine learning algorithms using artificial neural networks that have been developed so far are required to have selected features to learn in advance. Automated feature learning is one of the major characteristics of deep learning. Therefore, deep learning can be defined as a subset of machine learning in artificial intelligence with artificial neural networks that are able to learn without supervision from data and is also known as deep neural learning and deep neural network.

More intuitively, it is comparable to shallow learning. Let's assume that one trains a dog or other animals. Shallow learning is just direct intuitive learning by leading its action intentionally, for example, sitting down by pointing a finger and calling. In contrast, deep learning is more like learning a trend or behavior and making a decision by itself, for example, learning numbers by watching them. In order to make deep learning accessible, one must possess two important characteristics as remembering and learning from watching. Therefore, these two deep learning characteristics are defined in this book as temporal deep learning as remembering and spatial deep learning as learning from watching.

One of the major temporal deep learning techniques is recurrent neural network (RNN) models such as Long Short-term Memory and Gated Recurrent Unit. Also,

T. Lee et al., *Deep Learning for Hydrometeorology and Environmental Science*,
Water Science and Technology Library 99,
https://doi.org/10.1007/978-3-030-64777-3_1

Convolutional Neural Network is one of the major spatial deep learning techniques. In this book, these two deep learning techniques are mainly introduced, followed by applications of these deep learning techniques to hydrometeorological and environmental studies. Since those deep learning techniques are based on neural network models, the description of neural network models is given in advance.

1.2 Pros and Cons of Deep Learning

Deep learning has been recognized as the technology that makes artificial intelligence eventually become smart. Deep learning allows prediction by reducing the effort to find feature variables that are mostly time-consuming. With enough amount of data, not much human intervention is needed by deep learning to outperform other models. In other words, it can learn by itself from mimicking a human brain, especially in many layers of neurons in the brain cortex with the given dataset.

Deep learning is still pricy and resource-consuming and also requires a large amount of data. Since it is a branch of neural networks, it still lacks a strong theoretical foundation and the output result cannot generally provide theoretical reasoning and explanation. This unexplainable theoretical foundation and the requirement of a large dataset limits the application of deep learning models to hydrometeorology and environmental sciences.

1.3 Recent Applications of Deep Learning in Hydrometeorological and Environmental Studies

There are a number of recent developments and applications of deep learning models, especially to hydrometeorological and environmental studies. Some of the selected studies are discussed to present how deep learning models have been applied in the fields of hydrometeorology and environmental science.

In the hydrological field, temporal deep learning algorithms have also been applied in recent years. For example, Kratzert et al. (2018) applied a recent Recurrent Neural Network (RNN) model, named Long Short-Term Memory (LSTM), for rainfall-runoff modeling and compared it with Sacramento Soil Moisture Accounting Model (SAC-SMA) coupled with Snow-17 snow routine. Their results showed that LSTM had a competitive performance in comparison to the physical model of SAC-SMA, especially with regional scaling. Hu et al. (2018) compared artificial neural network (ANN) and LSTM for forecasting floods. Their results indicated that LSTM outperformed the conventional ANN and also showed that the model was more stable. Lee et al. (2020) proposed a stochastic simulation model with the LSTM model and their results indicated that the LSTM stochastic simulation model reproduced long-term variability as well as short-term memory.

In the meteorological field, Pan et al. (2019) employed the spatial deep learning model of Convolution Neural Network (CNN) to improve precipitation estimation in statistical downscaling. They trained the model to learn precipitation-related dynamical features from the surrounding dynamical fields and their results showed that the proposed model improved the precipitation-related parameterization scheme with CNN. Miao et al. (2019) applied the combined model of CNN and LSTM to forecast monsoon precipitation and compared it with ECMWF-Interim reanalysis precipitation. Their results showed that the combined deep learning model was superior to the physical model in forecasting precipitation more accurately from 1 day to 2 weeks in advance.

In environmental applications, Park et al. (2019) applied a deep neural network to model membrane fouling mechanisms and compared them with mathematical models. Their results indicated that the deep neural network showed better predictive performance in the fouling growth simulation and the flux decline simulation. Oga et al. (2019) applied CNN to estimate the water quality of a river. Using monitoring images, water quality can be estimated by training CNN, and it was observed that the proposed deep learning model outperformed the existing method in terms of accuracy.

1.4 Organization of Chapters

The chapters of this book are divided into three parts. The first part covers the introduction of the basic concepts of neural network models from Chaps. 1–7. Mathematical background and data preprocessing are given in Chaps. 2 and 3, respectively. Chapter 4 discusses the basic concept of a neural network and its training procedure is explained in Chap. 5. Chapter 6 explains how to update the weights of the neural network, followed by techniques to improve the model performance in Chap. 7.

The second part describes the advanced techniques in deep learning algorithms from Chaps. 8–10. In Chap. 8, advanced neural network algorithms, such as Extreme Learning Machine and Autoencoding, are discussed in Chap. 8. Chapter 9 deals with deep learning techniques, such as a recurrent neural network (RNN), long short-term memory (LSTM), and Gated Recurrent Unit (GRU), for time series data. Deep learning techniques for spatial datasets, as convolution neural network, are presented in Chap. 10.

The third part demonstrates the application procedure for deep learning techniques. Tensorflow and Keras programming are discussed in Chap. 11. Hydrometeorological and environmental applications to deep learning algorithms are presented in Chaps. 12 and 13, respectively.

1.5 Summary and Conclusion

Nowadays, there is a great deal of interest in the development and application of deep learning techniques in a number of fields. It is hoped that this book helps appreciate deep learning techniques, improve their understanding, and advance their application to hydrometeorological and environmental fields. Undergraduate and graduate students who are interested in deep learning algorithms in hydrometeorological and environmental fields may find the book to be useful. For beginners in governmental and educational sectors who need to apply deep learning techniques, this book might serve as a guide to become familiar with computational procedures for deep learning.

References

Hu C et al (2018) Deep learning with a long short-term memory networks approach for rainfall-runoff simulation. Water (Switzerland), 10(11). https://doi.org/10.3390/w10111543

Kratzert F, Klotz D, Brenner C, Schulz K, Herrnegger M (2018) Rainfall-runoff modelling using Long Short-Term Memory (LSTM) networks. Hydrol Earth Syst Sci 22(11): https://doi.org/10.5194/hess-22-6005-2018

Lee T, Shin JY, Kim JS, Singh VP (2020) Stochastic simulation on reproducing long-term memory of hydroclimatological variables using deep learning model. J Hydrol 582. https://doi.org/10.1016/j.jhydrol.2019.124540

Miao Q, Pan B, Wang H, Hsu K, Sorooshian S (2019) Improving monsoon precipitation prediction using combined convolutional and long short term memory neural network. Water (Switzerland) 11(5). DOI:https://doi.org/10.3390/w11050977

Oga T, Umeki Y, Iwahashi M, Matsuda Y (2019) River water quality estimation based on convolutional neural network, 2018 Asia-Pacific Signal and Information Processing Association Annual Summit and Conference, APSIPA ASC 2018—Proceedings, pp 1305–1308. https://doi.org/10.23919/APSIPA.2018.8659554

Pan B, Hsu K, AghaKouchak A, Sorooshian S (2019) Improving precipitation estimation using convolutional neural network. Water Resour Res 55(3): https://doi.org/10.1029/2018WR024090

Park S et al (2019) Deep neural networks for modeling fouling growth and flux decline during NF/RO membrane filtration. J Membrane Sci 587. https://doi.org/10.1016/j.memsci.2019.06.004

Chapter 2
Mathematical Background

Abstract In this current chapter, the fundamental mathematical background is presented for a deep learning model. Linear simple and multiple regression models are explained, including the definition of error terms and parameter estimation procedure, since they are similarly used in deep learning models. Also, the basic concept of the time series model is also explained and this part is mainly referred to in the LSTM model chapter.

In the current chapter, the fundamental mathematical background, including linear regression and time series model, is presented.

2.1 Linear Regression Model

2.1.1 Simple Linear Regression

A simple linear model can be described as $y = \beta_0 + \beta_1 x$ with the actual observed value of y as a linear function of x with parameters β_0 and β_1. Here, x is called as a predictor, explanatory variable, independent variable or input variable, while y is as predictand, response variable, dependent variable or output variable. This linear model can be generalized to a probabilistic model for the random variable Y as

$$Y = \beta_0 + \beta_1 X + \varepsilon \tag{2.1}$$

Note that Y is capitalized because the output of the model includes a random noise (ε) assumed to be normally distributed with mean $E(\varepsilon) = 0$ and $Var(\varepsilon) = \sigma_\varepsilon^2$ and the output is now a random variable. Also, X is called a predictor (or an explanatory variable), while Y is a predictand (or a response variable).

T. Lee et al., *Deep Learning for Hydrometeorology and Environmental Science*,
Water Science and Technology Library 99,
https://doi.org/10.1007/978-3-030-64777-3_2

For a sample of observed data pairs of size n, that is, (x_i, y_i) for $i = 1, \ldots, N$, the sum of squares of errors (SSE) is defined as

$$SSE = \sum_{i=1}^{N} [y_i - \hat{y}_i]^2 = \sum_{i=1}^{N} [y_i - (\beta_0 + \beta_1 x_i)]^2 = \sum_{i=1}^{N} \varepsilon_i^2 \tag{2.2}$$

The parameters can be estimated by finding the parameter set with the criterion minimizing the sum of errors (i.e., SSE), called least-square estimate, which is analogous to taking the derivatives of SSE with respect to the parameters separately and equating the derivatives to zero as

$$\frac{\partial SSE}{\partial \beta_0} = -2\sum_{i=1}^{N} [y_i - (\beta_0 + \beta_1 x_i)] = 0 \tag{2.3}$$

$$\frac{\partial SSE}{\partial \beta_1} = -2\sum_{i=1}^{N} [y_i - (\beta_0 + \beta_1 x_i)](x_i) = 0 \tag{2.4}$$

From Eq. (2.3),

$$\sum_{i=1}^{N} y_i - n\beta_0 - \beta_1 \sum_{i=1}^{n} x_i = 0 \tag{2.5}$$

$$\hat{\beta}_0 = \overline{y} - \hat{\beta}_1 \overline{x} \tag{2.6}$$

From Eq. (2.4),

$$\sum_{i=1}^{N} [y_i - (\overline{y} - \beta_1 \overline{x} + \beta_1 x_i)] x_i = 0 \tag{2.7}$$

$$\sum_{i=1}^{N} [(y_i - \overline{y}) - \beta_1 (x_i - \overline{x})] x_i = 0 \tag{2.8}$$

Since $\sum_{i=1}^{N} [(y_i - \overline{y}) - \beta_1 (x_i - \overline{x})]\overline{x} = 0$,

$$\sum_{i=1}^{N} [(y_i - \overline{y}) - \beta_1 (x_i - \overline{x})] x_i - \sum_{i=1}^{N} [(y_i - \overline{y}) - \beta_1 (x_i - \overline{x})]\overline{x} = 0 \tag{2.9}$$

$$\sum_{i=1}^{N} [(y_i - \overline{y}) - \beta_1 (x_i - \overline{x})](x_i - \overline{x}) = 0 \tag{2.10}$$

Then,

$$\hat{\beta}_1 = \frac{\sum_{i=1}^{N} (x_i - \overline{x})(y_i - \overline{y})}{\sum_{i=1}^{N} (x_i - \overline{x})^2} \tag{2.11}$$

Note that the least-square estimate of β_0 and β_1 is now denoted as $\hat{\beta}_0$ and $\hat{\beta}_1$.

In addition, the coefficient of determination, the proportion of the variance in the predictand variable that is predictable from a predictor, is denoted as R^2 and can be estimated as follows:

$$R^2 = \frac{SSR}{SST} = \frac{\sum_{i=1}^{N}(\hat{y}_i - \overline{y})^2}{\sum_{i=1}^{N}(y_i - \overline{y})^2} \quad (2.12)$$

where the sum of squares of the total (SST) is $\sum_{i=1}^{n}(y_i - \overline{y})^2$ and the regression sum of squares (SSR) is $\sum_{i=1}^{N}(\hat{y}_i - \overline{y})^2$. This statistic is a measure of how well the predictor observations can be replaced by the model predictions according to the portion of the variation, as in Eq. (2.12). A higher value of R2 indicates better performance of the linear regression model.

Example 2.1 Estimate the parameters of β_0 and β_1 with the least-square estimate as shown in Eqs. (2.6) and (2.11), respectively, for the dataset in the second column (i.e., x_1) for a predictor (x) and the fourth column for a predictand (y) in Table 2.1.

Solution:

As shown in Table 2.2,

$$\hat{\beta}_1 = \frac{\sum_{i=1}^{N}(x_i - \overline{x})(y_i - \overline{y})}{\sum_{i=1}^{N}(x_i - \overline{x})^2} = \frac{1.88}{3.65} = 0.51$$

$$\hat{\beta}_0 = \overline{y} - \hat{\beta}_1\overline{x} = -0.62 - 0.51 \times 0.51 = -0.8$$

Its scatterplot is shown in Fig. 2.1 with its estimated equation. Note that the line indicates the estimated value from $\hat{y} = \hat{\beta}_0 + \hat{\beta}_1 x = -0.8 + 0.51x$.

Also, the coefficient of determination (R^2) can be estimated from Table 2.2 as

$$R^2 = \frac{SSE}{SST} = \frac{\sum_{i=1}^{n}(\hat{y}_i - \overline{y})^2}{\sum_{i=1}^{n}(y_i - \overline{y})^2} = \frac{0.97}{2.77} = 0.35$$

2.1.2 Multiple Linear Regression

The multiple regression for multiple variables of ${x_i}^1, {x_i}^2, \ldots, {x_i}^S$ can be described as

Table 2.1 Example dataset for simple and multiple linear regression

Index	$\mathbf{x}_1$	$\mathbf{x}_2$	Y
1	−0.33	−0.49	−1.83
2	0.69	0.25	−0.23
3	0.26	−0.69	−1.25
4	−0.22	−0.54	−0.70
5	0.50	0.16	0.03
6	0.28	−0.16	−0.31
7	1.76	−0.18	−0.23
8	0.35	0.45	−0.79
9	0.47	0.49	−0.49
10	1.33	0.42	−0.44
Average	0.51	−0.03	−0.62

Table 2.2 Simple linear regression example to estimate the parameters and determination of coefficient (R^2)

Index	$x_i - \overline{x}$	$y_i - \overline{y}$	$(x_i - \overline{x}) \times (y_i - \overline{y})$	$(x_i - \overline{x})^2$	$\hat{y}_i$	$\hat{y}_i - \overline{y}$	$(\hat{y}_i - \overline{y})^2$	$(y_i - \overline{y})^2$
1	−0.84	−1.21	1.01	0.70	−1.06	−0.43	0.19	1.45
2	0.18	0.39	0.07	0.03	−0.53	0.09	0.01	0.16
3	−0.25	−0.63	0.16	0.06	−0.75	−0.13	0.02	0.39
4	−0.73	−0.08	0.06	0.53	−1.00	−0.38	0.14	0.01
5	−0.01	0.65	−0.01	0.00	−0.63	0.00	0.00	0.43
6	−0.23	0.31	−0.07	0.05	−0.74	−0.12	0.01	0.10
7	1.25	0.39	0.49	1.57	0.02	0.65	0.42	0.16
8	−0.16	−0.17	0.03	0.03	−0.71	−0.08	0.01	0.03
9	−0.04	0.13	−0.01	0.00	−0.64	−0.02	0.00	0.02
10	0.82	0.18	0.15	0.67	−0.20	0.42	0.18	0.03
		Sum	**1.88**	**3.65**		Sum	**0.97**	**2.77**

$$Y_i = \beta_0 + \beta_1 x_i^1 + \beta_2 x_i^2 + \cdots + \beta_s x_i^s + \varepsilon_i \tag{2.13}$$

where S is the number of predictors of interest. In matrix form, the linear regression model can be expressed by

$$Y_i = \vec{\mathbf{x}}_i^T \boldsymbol{\beta} + \varepsilon_i \tag{2.14}$$

where $\overrightarrow{\mathbf{x}}_i = [1, x_i^1, x_i^2, \ldots, x_i^S]^T$ and $\boldsymbol{\beta} = [\beta_0, \beta_1, .., \beta_s]^T$. Note that 1 is added in $\overrightarrow{\mathbf{x}}_i$ to include the intercept term of β_0 in this matrix form.

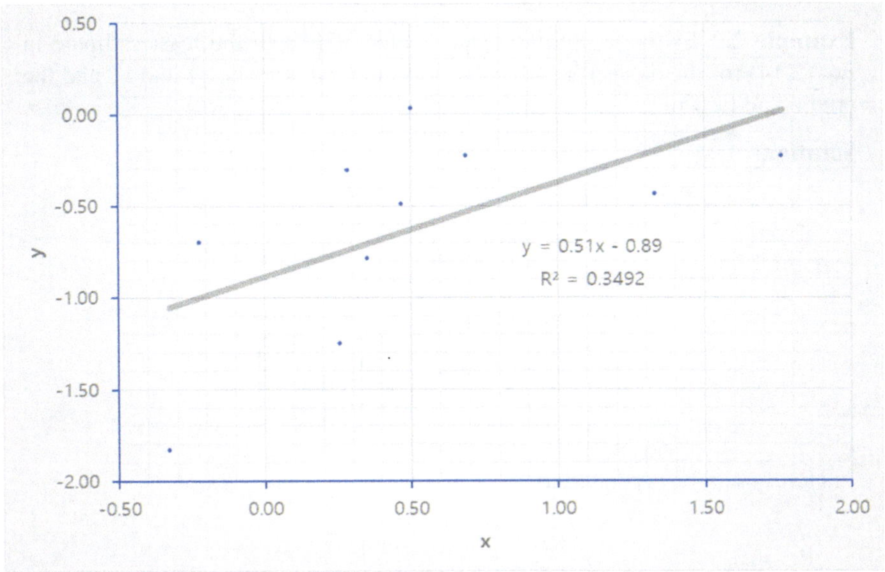

Fig. 2.1 Scatterplot (dots) and the fitted simple regression (line and equation) for Exercise 2.1

The sum of squares of residuals and its least-square estimate of $\boldsymbol{\beta}$ can be given by

$$SSE = \sum_{t=1}^{n} (\varepsilon_t)^2 = \mathbf{e}'\mathbf{e} = (\mathbf{y} - \overrightarrow{\mathbf{x}}\boldsymbol{\beta})^T(\mathbf{y} - \overrightarrow{\mathbf{x}}\boldsymbol{\beta})$$
$$= \mathbf{y}^{\mathrm{T}}\mathbf{y} - \mathbf{y}^{\mathrm{T}}\overrightarrow{\mathbf{x}}\boldsymbol{\beta} - \boldsymbol{\beta}^{\mathrm{T}}\overrightarrow{\mathbf{x}}^{\mathrm{T}}\mathbf{y} + \boldsymbol{\beta}^{\mathrm{T}}\overrightarrow{\mathbf{x}}^{\mathrm{T}}\overrightarrow{\mathbf{x}}\boldsymbol{\beta}$$

where $\mathbf{e} = [\varepsilon_1, \ldots, \varepsilon_n]^T$, $\overrightarrow{\mathbf{x}} = [1, \ldots, 1; {x_1}^1, \ldots., {x_n}^1; {x_1}^2, \ldots., {x_n}^2; \ldots.; {x_1}^S, \ldots., {x_n}^S]^T$ ($n \times S$+1 matrix) and $\mathbf{y} = [y_1, \ldots, y_n]^T$ with n number of data. The derivative to have its minimum with $dSS/d\boldsymbol{\beta} = 0$ can be obtained as Eq. (2.15):

$$\frac{dSSE}{d\boldsymbol{\beta}} = -2\overrightarrow{\mathbf{x}}^{\mathrm{T}}\mathbf{y} + 2\overrightarrow{\mathbf{x}}^{\mathrm{T}}\overrightarrow{\mathbf{x}}\boldsymbol{\beta} = 0$$

The least-square estimator can be obtained as

$$\hat{\boldsymbol{\beta}} = (\overrightarrow{\mathbf{x}}^{\mathrm{T}}\overrightarrow{\mathbf{x}})^{-1}\overrightarrow{\mathbf{x}}^{\mathrm{T}}\mathbf{y} \tag{2.15}$$

Also, the error variance $Var(\varepsilon) = \sigma_\varepsilon^2$ can be estimated with

$$\hat{\sigma}_\varepsilon^2 = \frac{SSE}{n - (k+1)} \tag{2.16}$$

where k is the number of parameters used.

Example 2.2 Estimate the parameter vector of $\boldsymbol{\beta}$ with the least estimate in Eq. (2.14) for the dataset in Table 2.1. Let the predictors be $x1$ and $x2$ and the predictand be y in Table 2.1.

Solution:

$$\vec{\mathbf{x}} = \begin{bmatrix} 1 & -0.33 & -0.49 \\ \vdots & \vdots & \vdots \\ 1 & 0.51 & -0.03 \end{bmatrix} \text{ and } \mathbf{y} = \begin{bmatrix} -1.83 \\ \vdots \\ -0.44 \end{bmatrix}$$

$$\hat{\boldsymbol{\beta}} = (\vec{\mathbf{x}}^{\mathrm{T}}\vec{\mathbf{x}})^{-1}\vec{\mathbf{x}}^{\mathrm{T}}\mathbf{y} = \begin{bmatrix} -0.80 \\ 0.38 \\ 0.45 \end{bmatrix}$$

Here, y can be predicted as

$$\hat{\mathbf{y}} = \vec{\mathbf{x}}\hat{\boldsymbol{\beta}} = \begin{bmatrix} -1.15 \\ \vdots \\ -0.62 \end{bmatrix}$$

Furthermore,

$$\hat{\sigma}_{\varepsilon}^{2} = \frac{SSE}{n-(k+1)} = \frac{1.50}{10-(3+1)} = 0.21$$

2.2 Time Series Model

2.2.1 *Autoregressive Model (AR)*

The autoregressive model of order p (AR(p)) for a time series variable Y_t can be generally described as

$$Y_t = \sum\nolimits_{j=1}^{p} \phi_j Y_{t-j} + \varepsilon_t \tag{2.17}$$

where Y_t is the time-dependent series with zero mean and ε_t is the time-independent series and the random white noise as $\varepsilon_t \sim \mathrm{N}(0,{\sigma_\varepsilon}^2)$ and ϕ_j is the parameter. Note that the mean of Y_t is assumed to be zero, which can be simply done by subtracting the mean of the original time series out. The simple AR(1) model, whose order (p) is one for AR(p), can be expressed as

$$Y_t = \phi_1 Y_{t-1} + \varepsilon_t \tag{2.18}$$

To estimate the parameter estimation, several approaches exist, such as the method of moments (MOM) and maximum likelihood (Brockwell and Davis 2003). For MOM, one can multiply Y_{t-1} and take expectation in Eq. (2.18) as

$$E\left[Y_t Y_{t-1}\right] = \phi_1 E\left[Y_{t-1} Y_{t-1}\right] + E\left[\varepsilon_t Y_{t-1}\right] \tag{2.19}$$

Then,

$$\gamma_1 = \phi_1 \sigma_Y^2 \tag{2.20}$$

where γ_1 is the lag-1 covariance for the Y variable, that is, $E\left[Y_t Y_{t-1}\right]$ with zero mean for Y and $E\left[Y_t Y_{t-1}\right] = \gamma_1$, $E\left[Y_{t-1} Y_{t-1}\right] = \sigma_Y^2$, and $E\left[\varepsilon_t Y_{t-1}\right] = 0$. Note that $E\left[\varepsilon_t Y_{t-1}\right] = 0$, since no dependence between the current error term and the previous data exists.

The parameter ϕ_1 becomes

$$\phi_1 = \rho_1 \tag{2.21}$$

where ρ_1 is the lag-1 correlation coefficient of Y variable, that is, $\rho_1 = \frac{E\left[Y_t Y_{t-1}\right]}{var(Y_t)}$.

By multiplying Y_t and taking expectation in Eq. (2.18), the variance of the error term $\left(\sigma_\varepsilon^2 = E[\varepsilon_t \varepsilon_t]\right)$ can be estimated as

$$E[Y_t Y_t] = \phi_1 E\left[Y_{t-1} Y_t\right] + E[\varepsilon_t Y_t] \tag{2.22}$$

$$E[Y_t Y_t] = \phi_1 E\left[Y_{t-1} Y_t\right] + E\left[\varepsilon_t (\phi_1 Y_{t-1} + \varepsilon_t)\right] \tag{2.23}$$

$$\sigma_Y^2 = \phi_1 \gamma_1 + E[\varepsilon_t \varepsilon_t] \tag{2.24}$$

$$\sigma_Y^2 = \phi_1 \gamma_1 + \sigma_\varepsilon^2 \tag{2.25}$$

Therefore,

$$\sigma_\varepsilon^2 = \sigma_Y^2 - \phi_1 \gamma_1 = \sigma_Y^2 (1 - \phi_1 \rho_1) = \sigma_Y^2 \left(1 - \phi_1^2\right)$$

Also, by multiplying $Y_{t\text{-}k}$ and taking expectation in Eq. (2.18)

$$E\left[Y_t Y_{t-k}\right] = \phi_1 E\left[Y_{t-1} Y_{t-k}\right] + E\left[\varepsilon_t Y_{t-k}\right] \tag{2.26}$$

$$\rho_k = \phi_1 \rho_{k-1} = \phi_1^k \tag{2.27}$$

Note that Eq. (2.27) shows that the lagged correlations are exponentially decreasing along with time lag k, implying that no long-term memory exists. The infinite sum $\sum_{k=-\infty}^{\infty} |\rho(k)| < \infty$ is defined as short-term memory and the case $\sum_{k=-\infty}^{\infty} |\rho(k)| = \infty$ is long-term memory (Brockwell and Davis 1988). Generally, the AR model has a short-memory structure.

Example 2.3 Estimate the parameters (ϕ_1 and σ_ε^2) of the AR(1) model for the data set in Table 2.3, obtained from the Colorado River streamflow at Lees Ferry, Colorado.

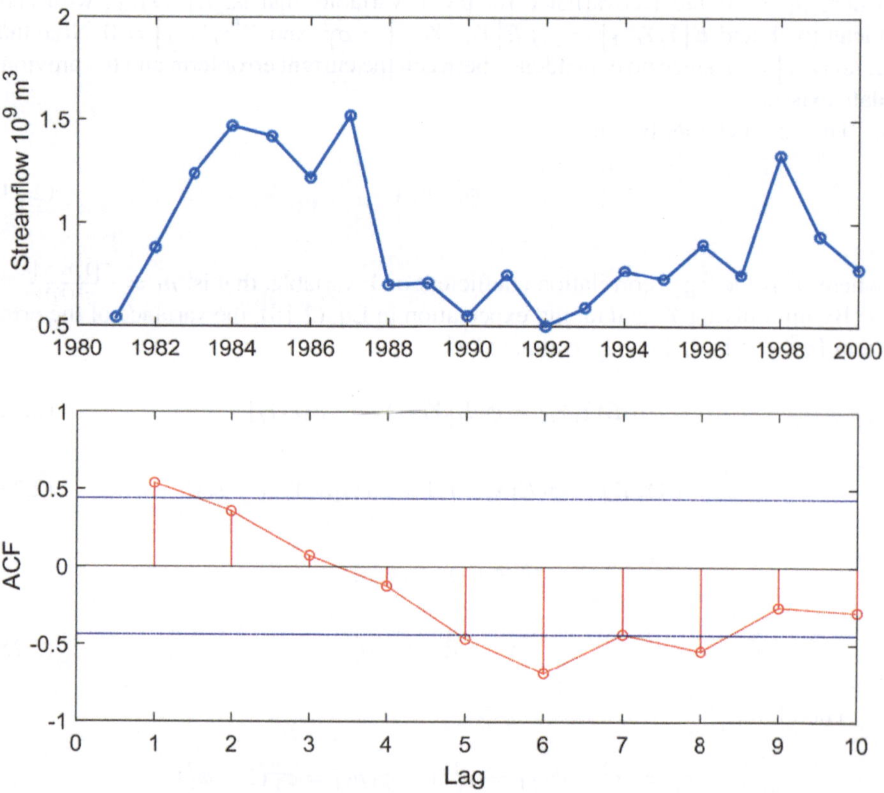

Fig. 2.2 Time series and autocorrelation functions of the Colorado River streamflow data in Table 2.3

Table 2.3 Estimation of the lag-1 autoregression model (AR-1) for the Colorado River streamflow at Lees Ferry

Ind	Year	Ann(10^9m^3)	y_t	y_{t-1}	y_t*y_{t-1}
1	1981	0.54	−0.37	−0.04	0.01
2	1982	0.88	−0.04	0.32	−0.01
3	1983	1.24	0.32	0.56	0.18
4	1984	1.47	0.56	0.50	0.28
5	1985	1.42	0.50	0.31	0.15
6	1986	1.22	0.31	0.60	0.19
7	1987	1.52	0.60	−0.22	−0.13
8	1988	0.70	−0.22	−0.20	0.04
9	1989	0.71	−0.20	−0.36	0.07
10	1990	0.55	−0.36	−0.17	0.06
11	1991	0.75	−0.17	−0.41	0.07
12	1992	0.50	−0.41	−0.32	0.13
13	1993	0.59	−0.32	−0.14	0.05
14	1994	0.77	−0.14	−0.18	0.03
15	1995	0.73	−0.18	−0.02	0.00
16	1996	0.90	−0.02	−0.16	0.00
17	1997	0.75	−0.16	0.42	−0.07
18	1998	1.33	0.42	0.03	0.01
19	1999	0.94	0.03	−0.13	0.00
20	2000	0.78	−0.13		
	Mean	0.914	0.000	**Sum**	**0.056**
	Var	0.109	0.103	ρ_1	**0.545**

Solution:

See the time series of streamflow shown in the upper panel of Fig. 2.2. The streamflow data (unit: 10^9m^3) in the third column of Table 2.3 is subtracted by mean and shown in the fourth column. The fourth and fifth columns are the same but with one step delayed as y_{t-1} and y_t.

$$\hat{\rho}_1 = \frac{E[Y_t Y_{t-1}]}{var(Y_t)} = \frac{E[Y_t Y_{t-1}]}{var(Y_t)} = \frac{\frac{1}{n-1}\sum_{t=1}^{n-1} y_t y_{t-1}}{\hat{\sigma}_Y^2} = \frac{0.056/19}{0.103} = 0.545$$

The lower panel of Fig. 2.2 illustrates the autocorrelation functions (ACFs) for k=1,…,10. In general, the ACF decreases with the lag since its correlation is weakened with further lag-time. Note that the number of data used for estimating $E[Y_t Y_{t-1}]$ is n−1 due to Y_{t-1}.Therefore, $\hat{\phi}_1 = \hat{\rho}_1 = 0.545$. Then, the error variance is

$$\hat{\sigma}_\varepsilon^2 = \hat{\sigma}_Y^2\left(1 - \hat{\phi}_1^2\right) = 0.103(1 - 0.545^2) = 0.072$$

$$\hat{\sigma}_\varepsilon = 0.269$$

2.3 Probability Distributions

2.3.1 Normal Distributions

The normal distribution, also known as Gaussian distribution, is the most widely employed probability distribution. Its PDF for a random variable X can be written as

$$f_X(x) = \frac{1}{\sigma\sqrt{2\pi}} \exp\left(-\frac{(x-\mu)^2}{2\sigma^2}\right) \quad -\infty < x < \infty \tag{2.28}$$

where μ and σ^2 are the parameters of this distribution represented by mean and variance. Note that mean μ locates the center of the normal distribution, while variance σ^2 shows its dispersion.

With the parameters of $\mu = 0$ and $\sigma = 1$, the distribution becomes the standard normal distribution. A variable with the normal distribution (i.e., $X \sim N(\mu, \sigma)$) can be transformed as $Z = \frac{X-\mu}{\sigma}$. Then, the variate Z becomes a standard normal, whose pdf can be written as

$$f_Z(z) = \frac{1}{\sqrt{2\pi}} \exp\left(-\frac{z^2}{2}\right) \tag{2.29}$$

Sometimes, it is also referred to as error distribution or function. The CDF of the standard normal distribution is

$$F_Z(z) = \int_{-\infty}^{z} \frac{1}{\sqrt{2\pi}} exp\left(-\frac{t^2}{2}\right) dt = \Phi(z) \tag{2.30}$$

The sample mean and sample variance for the parameters of normal distribution (μ and σ^2) are estimated with sample data of $x_1, \ldots, x_n$ as

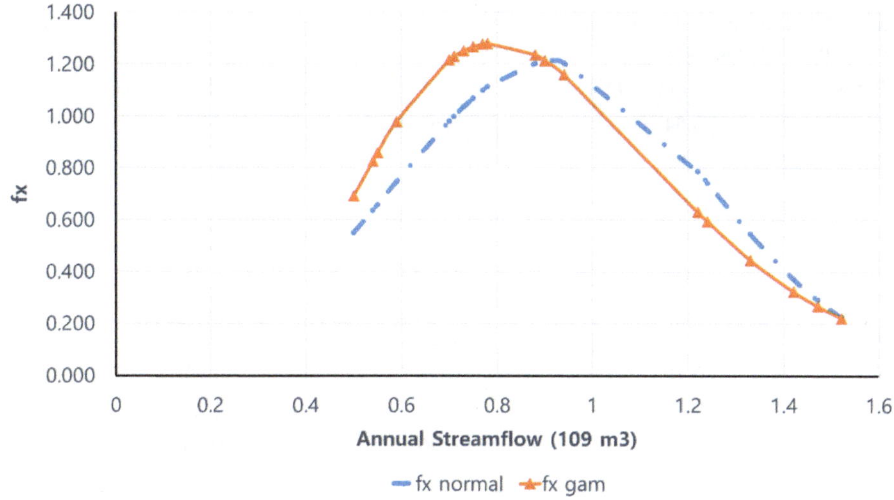

Fig. 2.3 Normal (blue dash-dotted line) and gamma (red solid line with triangle marker) probability density functions fitted to the Colorado River streamflow data in Table 2.3

$$\overline{x} = \frac{1}{n}\sum\nolimits_{i=1}^{n} x_i \tag{2.31}$$

$$\hat{\sigma}^2 = \frac{1}{n-1}\sum\nolimits_{i=1}^{n} (x_i - \overline{x})^2 \tag{2.32}$$

Example 2.4 Estimate the mean and variance of the data in Table 2.3, the annual Colorado River streamflow at Lees Ferry with Eqs. (2.31) and (2.32) and fit the data to the normal distribution.

Solution:

For the mean,

$$\overline{x} = \frac{1}{n}\sum_{i=1}^{n} x_i = \frac{1}{20}0.54 + 0.88 + \cdots + 0.78 = 0.91$$

$$\hat{\sigma}^2 = \frac{1}{n-1}\sum_{i=1}^{n}(x_i - \overline{x})^2 = \frac{1}{19}\left((0.54 - 0.914)^2 + \cdots + (0.75 - 0.914)^2\right) = 0.11$$

Table 2.4 Probability density functions of normal and gamma distributions for the annual Colorado River streamflow at Lees Ferry

Index	Ann($10^9 m^3$)	f(x) normal	f(x) gamma
1	0.54	0.64	0.82
2	0.88	1.20	1.24
3	1.24	0.74	0.59
4	1.47	0.29	0.27
5	1.42	0.37	0.32
6	1.22	0.79	0.63
7	1.52	0.22	0.22
8	0.70	0.98	1.22
9	0.71	1.00	1.23
10	0.55	0.66	0.86
11	0.75	1.07	1.27
12	0.50	0.55	0.69
13	0.59	0.75	0.98
14	0.77	1.10	1.28
15	0.73	1.03	1.25
16	0.90	1.21	1.21
17	0.75	1.07	1.27
18	1.33	0.55	0.44
19	0.94	1.20	1.16
20	0.78	1.11	1.28
Mean	**0.91**	**Alpa**	**7.67**
Var	**0.11**	**Beta**	**0.12**

The standard deviation is

$$\hat{\sigma} = \sqrt{0.109} = 0.33$$

For f_x of the normal distribution at x=0.54,

$$f_X(x) = \frac{1}{\hat{\sigma}\sqrt{2\pi}} exp\left(-\frac{(x-\overline{x})^2}{2\hat{\sigma}^2}\right) = \frac{1}{0.33\sqrt{2\pi}} exp\left(-\frac{(0.54-0.91)^2}{2\times 0.11}\right) = 0.64$$

The whole estimated pdf of the normal distribution is shownin Table 2.4 and Fig. 2.3. The plot shows that the peak of the normal pdf (f(x)) is 0.91, the same as the mean. Note that the normal distribution is symmetric in that the shape of the distribution is identical with the mean centered.

Note that by transforming with $Z = \frac{X-\mu}{\sigma}$, the pdf of the standard normal distribution

$$z = \frac{0.54 - 0.91}{0.33} = -1.12$$

$$fz(z) = \frac{1}{\sqrt{2\pi}} exp\left(-\frac{(-1.12)^2}{2}\right) = 0.21$$

2.3.2 *Gamma Distribution*

The gamma distribution has been widely used to fit hydrometeorological data, especially. Its probability density function (PDF) and cumulative distribution function (CDF) can be written as

$$f_G(x) = \frac{1}{\beta^\alpha \Gamma(\alpha)} x^{\alpha-1} \exp\left(-\frac{x}{\beta}\right) \tag{2.33}$$

$$F_G(x) = \int_0^x \frac{1}{\beta^\alpha \Gamma(\alpha)} x^{\alpha-1} \exp\left(-\frac{x}{\beta}\right) dx; \quad x, \alpha, \beta > 0 \tag{2.34}$$

where Γ(.) is the gamma function, and α and β are the shape and scale parameters, respectively. Its expectation and variance have the following relationship with these parameters:

$$E(X) = \mu = \int_0^\infty x f(x) dx f_G(x) = \alpha\beta \tag{2.35}$$

$$Var(X) = E[X - E(X)]^2 = \sigma^2 = \alpha\beta^2 \tag{2.36}$$

The parameters α and β are estimated using various methods, such as the method of moments (MOM) and maximum likelihood (ML). For example, MOM can estimate the parameters as

$$\beta = \sigma^2/\mu \tag{2.37}$$

$$\alpha = \mu/\beta \tag{2.38}$$

Example 2.5 Estimate the parameters of the gamma distribution in Table 2.3, the annual Colorado River streamflow at Lees Ferry with Eqs. (2.38) and (2.37) and fit the data to the gamma distribution.

Solution:

For the mean,

$$\hat{\beta} = \frac{\hat{\sigma}^2}{\overline{x}} = \frac{0.109}{0.914} = 0.12$$

$$\hat{\alpha} = \frac{\overline{x}}{\hat{\beta}} = \frac{0.914}{8.11} = 7.67$$

For f_x of the gamma distribution at x = 0.54,

$$fG(x) = \frac{1}{\hat{\beta}^{\hat{\alpha}}\Gamma(\hat{\alpha})}x^{\hat{\alpha}-1}exp\left(-\frac{x}{\hat{\beta}}\right) = \frac{1}{0.12^{7.67}\Gamma(7.67)}x^{0.12-1}exp\left(-\frac{0.54}{0.12}\right) = 0.82$$

The whole estimated pdf of the gamma distribution is shown in Table 2.4 and Fig. 2.3.

Note that the pdfs of the gamma and normal distributions are similar to each other, even though they have different statistical characteristics. The gamma distribution has a higher peak than the normal distribution. This is because the gamma distribution can have high skewness, while the normal distribution cannot be skewed with a symmetric shape.

2.4 Exercises

1. Estimate the mean and standard deviation for the following Colorado River data in Table 2.3, but only with the first 10 values (i.e., index 1−10).
2. Perform the simple linear regression with only the five values of the data in Table 2.1. Discuss how it is different from Example 1.1 and why.
3. Perform the multiple linear regression with only the five values of the data (i.e., index 1–5) in Table 2.1.
4. Derive the MOM for the parameters of the AR(2) model as below, employing the lagged correlation coefficients of ρ_1 and ρ_2.

$$Y_t = \phi_1 Y_{t-1} + \phi_2 Y_{t-2} + \varepsilon_t$$

5. Fit the AR(2) model to the dataset in Table 2.3.
6. Estimate the first moment and the probability density function of exponential distribution (i.e., $f_E(x) = \frac{1}{\lambda}\exp(-\frac{x}{\lambda})$) in Table 2.3.

References

Brockwell PJ, Davis R (1988) Simple consistent estimation of the coeffcients of a linear filter. Stochastic Process Appl 28:47–59

Brockwell PJ, Davis RA (2003) Introduction to time series and forecasting. Springer, Harrisonburg, VA, p 456

Chapter 3
Data Preprocessing

Abstract Before applying a neural network model, the data must be preprocessed in advance. Data normalization must be made to avoid the difference of data variability for inputs and outputs. It also simplifies the parameter range of a network model. Furthermore, data should be split for different purposes such as training, validation and testing. In this chapter, these data normalization and data split are explained in detail.

3.1 Normalization

In neural network, large ranges of input and output datasets pose difficulty for training the network. When there are a number of input or output variables and their scales are different from each other, then a variable with large variance might dominate the whole variables. Therefore, it is recommended to normalize input and output variables. One of the most common approaches is the minmax scaler, defined as

$$x_i = \frac{x_1^{\text{org}} - x_{\text{min}}^{\text{org}}}{x_{\text{max}}^{\text{org}} - x_{\text{min}}^{\text{org}}} \tag{3.1}$$

Also, standardization can be employed as

$$x_i = \frac{x_i^{\text{org}} - \mu_x^{\text{org}}}{\sigma_x^{\text{org}}} \tag{3.2}$$

where μ_x^{org} and σ_x^{org} are the mean and standard deviation of the original x variable.

The standardized dataset also needs to be back-normalized as follows. For minmax scaler,

$$x_i^{\text{org}} = x_i \left(x_{\text{max}}^{\text{org}} - x_{\text{min}}^{\text{org}} \right) + x_{\text{min}}^{\text{org}} \tag{3.3}$$

For standardization,

T. Lee et al., *Deep Learning for Hydrometeorology and Environmental Science*,
Water Science and Technology Library 99,
https://doi.org/10.1007/978-3-030-64777-3_3

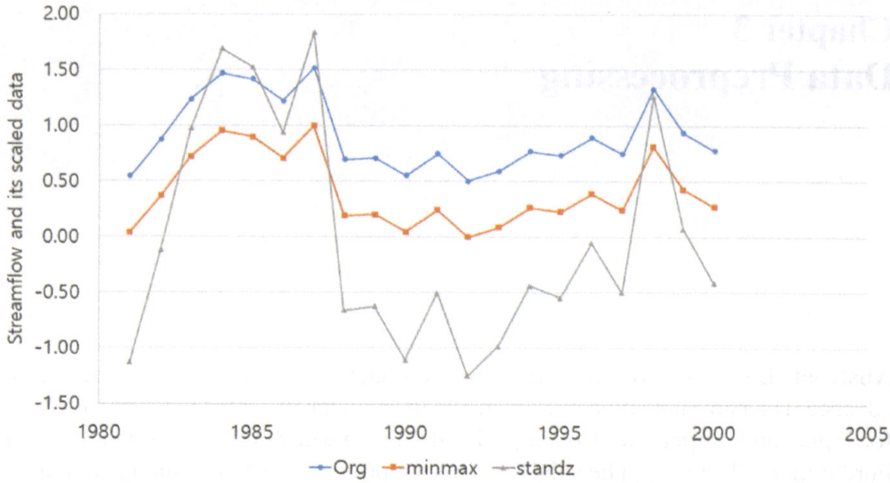

Fig. 3.1 Time series of original streamflow (109 m^3) and their minmax and standard scaled data for the Colorado River streamflow data in Table 2.3

$$x_i^{org} = x_i \sigma_{max}^{org} + \mu_{max}^{org} \tag{3.4}$$

Example 3.1 Normalize the dataset in Table 3.1 for the Colorado River streamflow at Lees Ferry.

Solution:

It is simple and direct. For the minmax scale,

$$x_i = \frac{x_i^{org} - x_{min}^{org}}{x_{max}^{org} - x_{min}^{org}} = \frac{0.54 - 0.50}{1.42 - 0.20} = 0.04$$

For standardization,

$$x_i = \frac{x_i^{org} - \mu_x^{org}}{\sigma_x^{org}} = \frac{0.54 - 0.91}{0.33} = -1.13$$

Table 3.1 and Fig. 3.1 show all the scaled data as well as the observation. The minmax scaler lowers the original data and have 0.0 and 1.0 values always. The standardization brings higher variability than has the original data, since the original variance is just $0.099 = 0.33^2$

See the time series of the streamflow shown in the upper panel of Fig. 2.2. The streamflow data (unit: $10^9 m^3$) in the third column of Table 2.3 is subtracted by mean and shown in the fourth column. The fourth and fifth column is the

Table 3.1 Observed data and its scaled data for the annual Colorado River streamflow at Lees Ferry

	Org	Minmax	Standz
1981	0.54	0.04	−1.13
1982	0.88	0.37	−0.11
1983	1.24	0.72	0.98
1984	1.47	0.95	1.69
1985	1.42	0.90	1.52
1986	1.22	0.71	0.94
1987	1.52	1.00	1.83
1988	0.70	0.19	−0.66
1989	0.71	0.20	−0.62
1990	0.55	0.05	−1.11
1991	0.75	0.24	−0.50
1992	0.50	0.00	−1.25
1993	0.59	0.09	−0.98
1994	0.77	0.26	−0.44
1995	0.73	0.23	−0.55
1996	0.90	0.39	−0.05
1997	0.75	0.24	−0.50
1998	1.33	0.82	1.27
1999	0.94	0.43	0.08
2000	0.78	0.27	−0.41
Mean	0.91		
Min	0.50		
Max	1.52		
Max-Min	1.01		
Std	0.33		

same but with one step delayed as y_{t-1} and y_t.

$$\rho_1 = \frac{E\left[Y_t Y_{t-1}\right]}{var(Y_t)} = \frac{E\left[Y_t Y_{t-1}\right]}{var(Y_t)} = \frac{\frac{1}{n-1}\sum_{t=1}^{n-1} y_t y_{t-1}}{\hat{\sigma}_{y_t}^2} = \frac{0.056/19}{0.103} = 0.545$$

Note that the number of data used for estimating $\left[Y_t Y_{t-1}\right]$, n−1 due to Y_{t-1}. The lower panel of Fig. 2.2 illustrates the autocorrelation functions (ACFs) for k = 1,…,10. In general, the ACFs decreases along with lags since its correlation is weakened with further lag-time.

One can program this scaling with Python. In this book, some examples of python program are provided, assuming that a reader has a little bit of experience in programming and Python.

```
[Python 4.01 ] Program the normalization procedure as minmax_scaler and stz_scaler

import numpy as np

def minmax_scaler(X):

  return (X-np.min(X))/(np.max(X)-np.min(X))

def stz_scaler(X):

  return (X-np.mean(X))/np.std(X)
```

In this simple example, numpy is employed to estimate minimum, maximum, mean, and standardization. Numpy is a package for scientific and statistical computing with N-dimensional objects, functions, algebra, and random numbers (https://numpy.org).

```
def minmax_scaler(X):

  return (X-np.min(X))/(np.max(X)-np.min(X))
```

This is how to define a function. The name of the function is "minmax_scaler" and its input data is "*x*."

3.2 Data Splitting for Training and Testing

In order to fit an ANN model, the model must be trained with observed data. Some part of the observed data must be held to provide an estimate of model skill to tune the hyperparameters of the model, such as the number of hidden units, the number of layers, and a type of activation function. This part of the observed dataset is used for validation. Furthermore, the error from the test dataset provides an unbiased estimate of model skill since the test dataset has not been employed in model fitting. Therefore, the observation dataset is split into three parts as training, validation, and test sets.

At first, a part (e.g., 20%) of the total observations are held for testing the dataset. Then, the remaining part is split into training and validation sets (e.g., 80% and 20%, respectively). Again, a training set is used to fit the ANN model parameters while a validation set provides an unbiased evaluation to tune hyperparameters, and a test set is used to evaluate the final model fit.

3.3 Exercises

1. Normalize the data in Table 2.1 with minmax scaler.
2. Normalize the data in Table 3.1 with the cumulative distribution function as

$$x_i = F_z^{-1}\left(F_G\left(x_i^{org}; \alpha, \beta\right)\right)$$

Note that F_Z and F_G are the cdfs of standard normal and gamma distribution, respectively.

Chapter 4
Neural Network

Abstract Deep learning model is composed of several layers of neural networks. Therefore, the basic concepts and terminology of a neural network are introduced. The simplest neural network model is introduced and used in the latter part of this book. Then, a full neural network model is described. The parameter estimation procedure employing backward propagation is also explained.

4.1 Terminology in Neural Network

4.1.1 Components of Neural Network

A neural network is made with a number of interconnected nodes, also known as neurons. A simple neural network contains three layers as input, hidden, and output with many neurons, as shown in Fig. 4.1. In this network, the source nodes of the input layer are projected onto the computational nodes in the hidden layer through the connection weights and then to the neurons of the output layer. The nodes of the input layer correspond to the number of features or target variables that are wished to be fed into the network. Two input nodes are used in the network shown in Fig. 4.1. The number of output nodes corresponds to the number of variables to be predicted as an example of one node in Fig. 4.1. The hidden layer nodes perform a nonlinear transformation on the input layer nodes. The number of hidden nodes in a network is a common concern related to the complexity of the relationship between target input and output datasets, as an example of three nodes in Fig. 4.1. One can increase the number of hidden layers to develop further complex relations between input and output layers with a full grouping of and connecting hidden nodes. Additional hidden layers often require an extensive training set to compute weights for a network.

Bias nodes are added in a network presented with b circle in Fig. 4.1 to allow a shift in the output of an activation function, which will be explained in the following. The learning process of a neural network is to adjust weights to minimize the difference between the values predicted by the model and real data. It illustrates an example of a neural network with an input layer, a hidden layer, and an output layer.

T. Lee et al., *Deep Learning for Hydrometeorology and Environmental Science*, Water Science and Technology Library 99,
https://doi.org/10.1007/978-3-030-64777-3_4

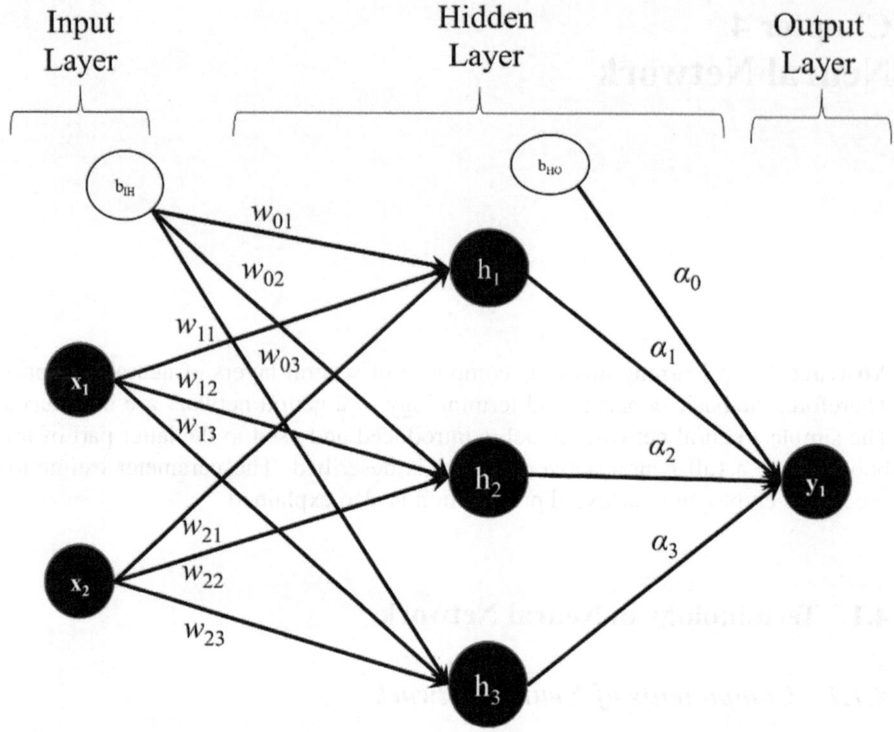

Fig. 4.1 Basic neural network with two inputs, one output, and three hidden nodes

4.1.2 Activation Functions

The activation function produces an output from each neuron bounding for the output neurons. Its selection is important to affect how to format input data. The most common activation functions are described in the following.

Linear function:
A linear function, presented as in Fig. 4.2, is one of the most basic activation functions.

$$h(s) = s \tag{4.1}$$

It is the identity mapping and results in a linear regression combining with weights and a bias term (i.e., $s = wx + b$). With this linear activation function, nonlinearity cannot be modeled. However, it is often used in the output layer for a network that predicts a quantity (i.e., regression).

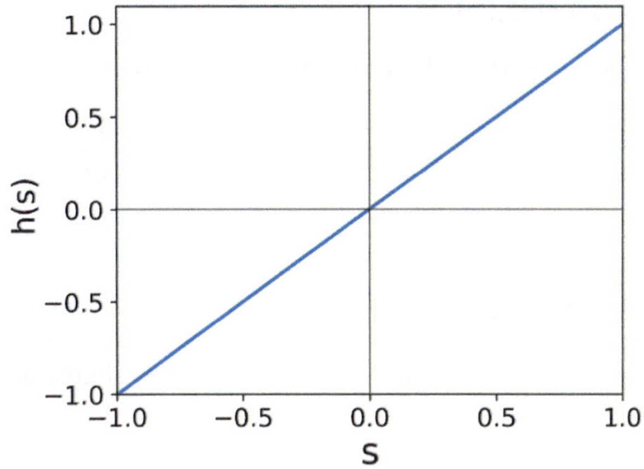

Fig. 4.2 Linear function for an activation function of a neural network

Step function:

The step function shown in Fig. 4.3 has been employed in earlier studies, called a perceptron. Its mathematical expression is

$$h(s) = \begin{cases} 1 & s > 0 \\ 0 & s \leq 0 \end{cases} \tag{4.2}$$

It is useful for problems with two outputs, such as a dry day or wet day, true or false, and so on. The threshold can be changed instead of the zero value as well as the magnitude before and after the threshold instead of zero and one (see Eq. (4.2)). Note that the step function is discontinuous at $x = 0$. This discontinuity hinders its application since continuous derivative information is needed overall the input information for learning by the neural network.

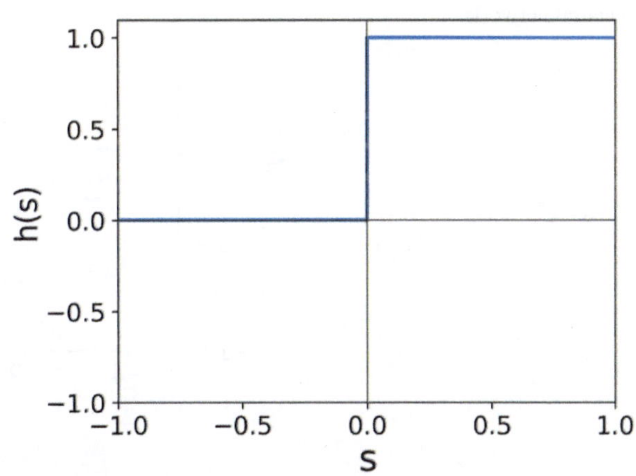

Fig. 4.3 Step function for an activation function of a neural network

Sigmoid function:

The sigmoid function can be simply denoted as

$$h(s) = \frac{1}{1 + \exp(-s)} \tag{4.3}$$

The sigmoid function is one of the most popularly employed activation functions due to its characteristics as continuity overall domain and monotonic increase. As shown in Fig. 4.4, the function is continuous from $-\infty$ to $+\infty$ and increases along with x increasing (monotonical increase). Furthermore, the function is in the interval between zero and one. Its derivative can be simply described as

$$\frac{\partial h}{\partial s} = (1 + \exp(-s))^{-2} \exp(-s) = h(1 - h) \tag{4.4}$$

Hyperbolic tangent (tanh) function:

The hyperbolic tangent function can be denoted as

$$h(s) = \tanh(s) = \frac{e^s - e^{-s}}{e^s + e^{-s}} \tag{4.5}$$

With the sigmoid function, the hyperbolic tangent function is also one of the most common activation functions due to its characteristics as continuity overall domain and monotonic increase. The function is continuous from $-\infty$ to $+\infty$ and increases along with x increasing, as in Fig. 4.5. Furthermore, the function is in the interval $[-1, 1]$. Its derivative can be described as

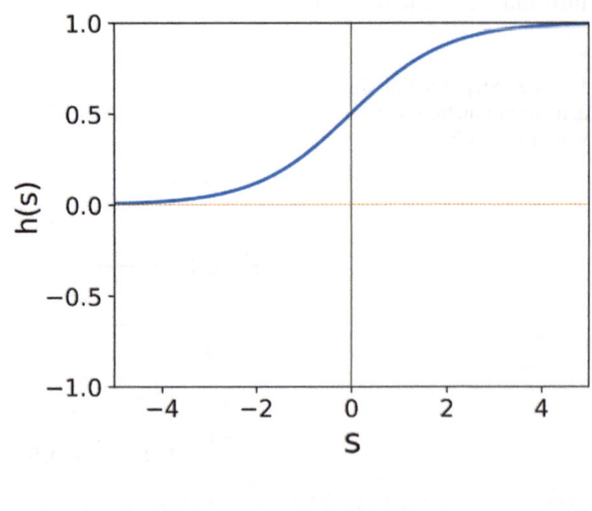

Fig. 4.4 Sigmoid function for an activation function of a neural network

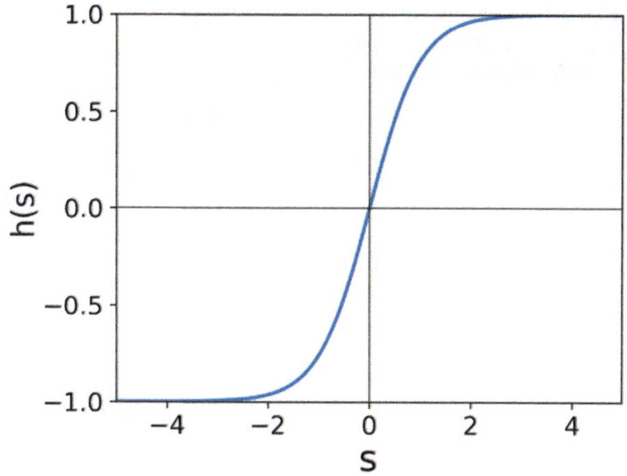

Fig. 4.5 Hyperbolic tangent (tanh) function for an activation function of a neural network

$$\frac{\partial h}{\partial s}=\frac{\left(e^{s}+e^{-s}\right)\left(e^{s}+e^{-s}\right)-\left(e^{s}-e^{-s}\right)\left(e^{s}-e^{-s}\right)}{\left(e^{s}+e^{-s}\right)^{2}}=1-\frac{\left(e^{s}-e^{-s}\right)^{2}}{\left(e^{s}+e^{-s}\right)^{2}}=1-h^{2} \tag{4.6}$$

ReLU function:

As shown in Fig. 4.6, the rectified linear unit (ReLU) function is described as

$$h(s)=\begin{cases} s & s>0 \\ 0 & s\leq 0 \end{cases} \tag{4.7}$$

or simply

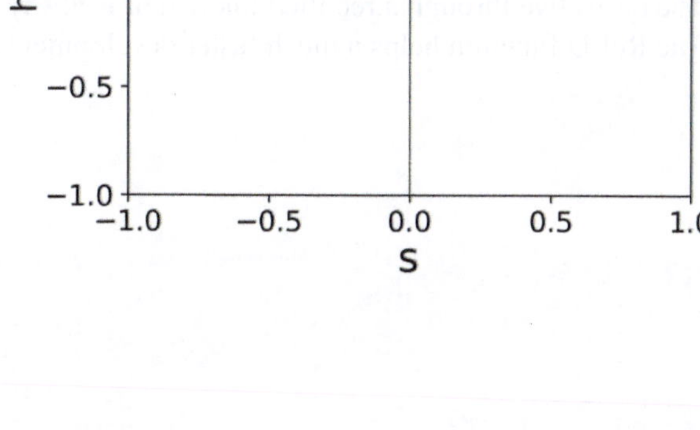

Fig. 4.6 Rectified Liner Unit (ReLU) for an activation function of a neural network

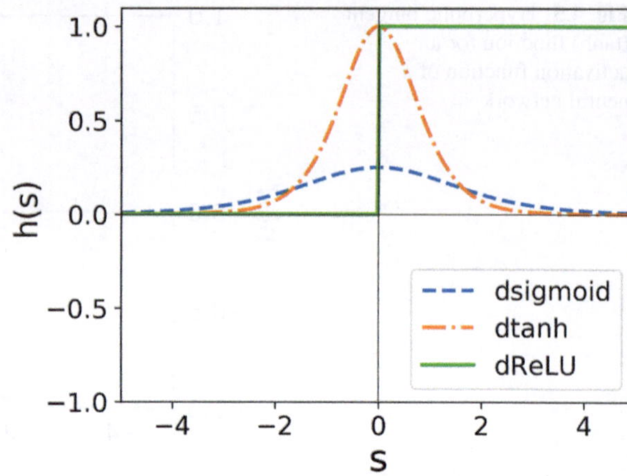

Fig. 4.7 Derivative of the popular activation functions as sigmoid, tanh, and ReLU

$$h(s) = \max\{0, s\} \tag{4.8}$$

The formula indicates that the ReLU function is the same as the linear activation function when s > 0. The only difference is that ReLU results in zero across half its domain. Its derivative can be described as

$$\frac{\partial h(s)}{\partial s} = \begin{cases} 1 \; s > 0 \\ 0 \; s < 0 \end{cases} \tag{4.9}$$

and undefined in $x = 0$.

Unlike the linear activation function, a nonlinear activation function allows a network learning more complex structures such as in sigmoid and tanh functions. However, these nonlinear functions have drawbacks, such as not much sensitive to the activation sum input and easily saturated in training. For example, Fig. 4.7 illustrates the derivative of three common activation functions. Sigmoid and tanh ($h(s)$) present close to zero at high or low s (see the left and right sides of the functions). Also, the derivative of the sigmoid is much flatter than its original function. Especially at $s = 0$ from Eq. (4.4), $h(0)(1 - h(0)) = 0.25$.

The linear part of ReLU inherits its good characteristics of linear activation function in training the neural network through backpropagation, while it is still a nonlinear function since negative values always result in zero, as shown in Fig. 4.7. The derivative through a rectified linear unit is always large when the unit is active. The ReLU function helps a much faster development of deep neural networks.

4.1.3 Error and Loss Function

In a prediction problem with one input and one output like in Eq. 2.1, one can simply describe it as

$$Y = f(x) + \varepsilon \tag{4.10}$$

where $f(x)$ is nonlinearly modeled with a neural network. The error is defined as

$$\varepsilon_i = y_{o,i} - f(x_i) = y_{o,i} - \hat{y}_{p,i} \tag{4.11}$$

where $y_{0,i}$ is the ith observed output value, while $\hat{y}_{p,i}$ is the corresponding estimated value. The hat in $\hat{y}_{p,i}$ indicates the estimate. From now onwards, $\hat{y}_{p,i}$ is simply presented as $y_{p,i}$.

The loss function is employed to measure the difference between a predicted value ($\hat{y}$) and its corresponding observation (y). Generally, it consists of the empirical risk term and a regularization term. Here, the empirical risk term is considered, and the regularization term is explained in a later chapter. There are a number of loss functions, such as L2, mean square error, mean absolute error, and cross-entropy.

The L2 loss function is the square of the L2 norm of the difference as in Eq. (4.11) and can be described as

$$L2 = \sum_{i=1}^{N} \left[y_{o,i} - y_{p,i}\right]^2 \tag{4.12}$$

where N is the number of observations.

Another loss function is the mean square error, dividing the SSE in Eq. 2.2with the number of records (n) defined as:

$$\text{MSE} = \frac{1}{n}\sum_{i=1}^{N} \left[y_{\text{o},i} - y_{p,i}\right]^2 \tag{4.13}$$

Also, cross-entropy error (CEE) has been popularly employed in classification as

$$\text{CEE} = -\sum_{i=1}^{N} y_{o,i} \log\left(y_{p,i}\right) \tag{4.14}$$

In general, $y_{o,i}$ is zero or one, while $y_{p,i}$is a probability from zero to one in a classification problem. This CEE originated from the informational entropy theory.

Informational entropy is defined as $I = -\log(p)$ by measuring the amount of uncertainty, that is, the lower the probability (here, p), the higher the information is. Lower probability indicates higher quality information since the value x does not occur often. The entropy (H), the measurement of uncertainty, refers to the expected value of the information (I) as $\mathrm{E}[I] = E[-\log(p)]$. For sampled data, the entropy (Shannon 1948) can be estimated as

$$H = -\sum_{i=1}^{n} p_i \log(p_i) \tag{4.15}$$

Note that $E(X) = \sum_{i=1}^{n} p_i x_i$ for a discrete variable X.

To derive CEE, the Kullback–Leibler (KL) divergence must be introduced as

$$KL(p|q) = -\sum_{i=1}^{N} p_i \log(q_i) - \left(-\sum_{i=1}^{N} p_i \log(p_i)\right) \tag{4.16}$$

where q_i is also the probability for the same variable x as p_i. the KL divergence is also called "Relative Entropy" and measures the difference of two probability distributions (p_i and q_i).

For CEE, assume that p_i is a target probability and q_i is a measured probability. The objective is now to update the parameters related to $q_i$so that q_i gets close to pi. The second term on the right side of Eq. (4.16) does not change according to $q_i$and does not contribute to finding the minimum of KL divergence. Therefore, by eliminating the second term, the KL divergence becomes CEE as $-\sum_{i=1}^{N} p_i \log(q_i)$ as in Eq. (4.14).

Example 4.1 Estimate CEE and MSE with the dataset (y_o and y_p) in Table 4.1

Solution:

$$\mathrm{CEE} = -\sum_{i=1}^{N} y_{0,i} \log(y_{p,i}) = -(0\log(0.3) + 0\log(0.3) + 1\log(0.4)\ldots) = 1.8$$

$$\mathrm{MSE} = \frac{1}{n}\sum_{i=1}^{N} \left[y_{0,i} - y_{p,i}\right]^2 = \frac{1}{9}\left[(0-0.3)^2 + (0-0.3)^2 + (1-0.4)^2 + \cdots\right] = 0.27$$

See Table 4.1 for detail.

CEE only considers the observed values in the target class (i.e., $y_{o,i} = 1$). Note that CEE is only focused on the values that are one, indicating the observed data is in the class, while MSE considers all the values.

Table 4.1 Example dataset for CEE and MSE in classification

Data						CEE				MSE			
y_p			y_o			$y_0 \log(y_p)$				$(y_0 - y_p)^z$			
Pr(A)	Pr(B)	Pr(c)	A	B	C	A	B	C	sum	A	B	C	sum
0.3	0.3	0.4	0	0	1	0.00	0.00	−0.4	−0.4	0.09	0.09	0.36	0.54
0.3	0.4	0.3	0	1	0	0.00	−0.4	0.00	−0.4	0.09	0.36	0.09	0.54
0.1	0.2	0.7	1	0	0	−1.0	0.00	0.00	−1.0	0.81	0.04	0.49	1.34
			Acc = 6/9 = 0.67					CEE	1.80			MSE	0.27

4.1.4 Softmax and One-Hot Encoding

Softmax and one-hot encoding algorithms should be applied for categorical data and classification problems.

One-hot Encoding:

One-hot encoding is applied to the categorical data (e.g., people, animal, and others in a picture) for classification. One-hot encoding is the most common approach for classification with a limited number of categories. In one-hot encoding, a binary matrix is made for the output categorical data.

For example, consider categorical data as human, animal, and others, and picture data is applied as input. Then, the binary matrix can be made for the output with one-hot encoding as

$$\begin{bmatrix} Human \\ Animal \\ Others \\ Animal \\ \vdots \end{bmatrix} = \begin{bmatrix} 1\,0\,0 \\ 0\,1\,0 \\ 0\,0\,1 \\ 0\,1\,0 \\ \vdots\,\vdots\,\vdots \end{bmatrix} \tag{4.17}$$

In the categorical data with true and false, one-hot encoding can be also used. For example, consider wet and dry days. The one-hot encoding binary matrix can be made as

$$\begin{bmatrix} Wet \\ Dry \\ Dry \\ Wet \\ \vdots \end{bmatrix} = \begin{bmatrix} 1\,0 \\ 0\,1 \\ 0\,1 \\ 1\,0 \\ \vdots\,\vdots \end{bmatrix} \tag{4.18}$$

Example 4.2 Make one-hot encoding for the precipitation data shown in Table 4.2.

Solution:

At first, the precipitation data in the fourth column of Table 4.2 is transferred to binary data (i.e., 0 and 1). So, whenever the amount is greater than 0 (wet day), it should be one. Otherwise, it should be zero as seen in the fifth column

Table 4.2 Example hourly temperature and precipitation dataset

Ind	Time	TA (°C)	PCP (mm)	PCP Bin
1	2,015,050,219	21.7	0	0
2	2,015,050,220	20.6	0	0
3	2,015,050,221	20.3	0	0
4	2,015,050,222	19.4	0.4	1
5	2,015,050,223	18.9	0.1	1
6	2,015,050,300	18.8	0	0
7	2,015,050,301	18.6	0	0
8	2,015,050,302	18	0.3	1
9	2,015,050,303	17.9	0.2	1
10	2,015,050,304	17.7	1	1
11	2,015,050,305	17.5	1	1
12	2,015,050,306	17.1	0	0
13	2,015,050,307	17.3	0	0
14	2,015,050,308	17.7	0	0
15	2,015,050,309	18.2	0	0
16	2,015,050,310	18.2	0.5	1
17	2,015,050,311	18.5	0	0
18	2,015,050,312	18.7	0	0
19	2,015,050,313	18.8	0	0
20	2,015,050,314	18.9	0	0

of Table 4.2. Note that there are two classes for precipitation, either 1 (wet) or 0 (dry).

For one-hot encoding, the one-hot encoding binary matrix is made as

$$\begin{bmatrix} 0 & 1 \\ 0 & 1 \\ 0 & 1 \\ 1 & 0 \\ \vdots & \vdots \end{bmatrix}$$

See the third and fourth columns of Table 4.2 for the one-hot encoding binary matrix.

Softmax:

Softmax is a function that normalizes an input vector (i.e., $N_{sm} \times 1$ where N_{sm} is the number of the target softmax variables) into a probability distribution proportional to the exponentials of each input value, denoted as:

$$y_p^c = \frac{\exp(z^c)}{\sum_{c'=1}^{N_{sm}} \exp\left(z^{c'}\right)} \tag{4.19}$$

where z^c is the input values of the cth variable, and y_p^c is the resulting predicted value of the cth variable.

This softmax function is commonly employed for classification in the final layer. In classification, a final output value must be with a probability with N_{sm} number of variables for N_{sm} classes and meet the probability characteristics as (1) the sum of probabilities of K variables must be one and (2) each value must be between zero and one. However, it is very difficult for an output variable from a neural network to meet these probability characteristics.

The additional softmax layer in Eq. (4.19) enables the layer to meet the conditions that: (1) the sum of the output variables is equal to one (i.e., $\sum_{c=1}^{N_{sm}} y_p^c = 1$), since each variable is already divided by the sum as $\sum_{c'=1}^{N_{sm}} \exp\left(z\wedge c'\right)$; and (2) each y_p^c value must be between zero and one (i.e., $0 \le y_p^c \le 1$) because the exponential is taken for the network output $(z^c)(z^c$ as $\exp(z^c)$ and divided by the sum. Note that the exponential function is an autonomically increasing function, and no negative value exists, implying that a softmax output value is always positive and does not have to be changed to the order of the original values.

Example 4.3 Perform the softmax procedure for the dataset of hourly temperature and precipitation in Table 4.2. Let the temperature be the predictor x and the precipitation occurrence (wet or dry) be the predictand y. Assume the weights and bias for the input and the softmax are the following:

$$\mathbf{W} = \begin{bmatrix} w_1 \\ w_0 \end{bmatrix} = \begin{bmatrix} -1.54 \\ 1.54 \end{bmatrix} \text{ and } \mathbf{b} = \begin{bmatrix} b_1 \\ b_0 \end{bmatrix} = \begin{bmatrix} 0.155 \\ -0.155 \end{bmatrix}$$

Solution:

$$z_t^1 = w_1 x_t + b_1 = -1.54 \times 1.00 + 0.155 = -1.39$$

$$z_t^2 = w_0 x_t + b_0 = 1.54 \times 1.00 - 0.155 = 1.39$$

Table 4.3 Softmax procedure for the temperature and precipitation dataset

	X	ONE	ZERO	Z_1	Z_2	EXP(Z1)	EXP(Z2)	SUM	Y_1	Y_0
1	1.00	0	1	−1.39	1.39	0.25	4.00	4.25	0.06	0.94
2	0.76	0	1	−1.02	1.02	0.36	2.76	3.13	0.12	0.88
3	0.70	0	1	−0.92	0.92	0.40	2.50	2.90	0.14	0.86
4	0.50	1	0	−0.61	0.61	0.54	1.85	2.39	0.23	0.77
5	0.39	1	0	−0.45	0.45	0.64	1.56	2.20	0.29	0.71
6	0.37	0	1	−0.41	0.41	0.66	1.51	2.17	0.30	0.70
7	0.33	0	1	−0.35	0.35	0.71	1.41	2.12	0.33	0.67
8	0.20	1	0	−0.15	0.15	0.86	1.16	2.02	0.43	0.57
9	0.17	1	0	−0.11	0.11	0.89	1.12	2.01	0.44	0.56
10	0.13	1	0	−0.05	0.05	0.96	1.05	2.00	0.48	0.52
11	0.09	1	0	0.02	−0.02	1.02	0.98	2.00	0.51	0.49
12	0.00	0	1	0.16	−0.16	1.17	0.86	2.02	0.58	0.42
13	0.04	0	1	0.09	−0.09	1.09	0.92	2.01	0.54	0.46
14	0.13	0	1	−0.05	0.05	0.96	1.05	2.00	0.48	0.52
15	0.24	0	1	−0.21	0.21	0.81	1.24	2.05	0.40	0.60
16	0.24	1	0	−0.21	0.21	0.81	1.24	2.05	0.40	0.60
17	0.30	0	1	−0.31	0.31	0.73	1.37	2.10	0.35	0.65
18	0.35	0	1	−0.38	0.38	0.68	1.46	2.15	0.32	0.68
19	0.37	0	1	−0.41	0.41	0.66	1.51	2.17	0.30	0.70
20	0.39	0	1	−0.45	0.45	0.64	1.56	2.20	0.29	0.71

Here, z_t^1 and z_t^2 are the linear transferred functions from input data for representing the wet and dry day classes, respectively. This estimation of the whole data is presented in the fifth and sixth columns of Table 4.3. From the softmax function in Eq. (4.19),

$$y_1^1 = \frac{\exp(z^1)}{\sum_{c'=1}^{N_{sm}} \exp(z^c)} = \frac{\exp(-1.39)}{\exp(-1.39) + \exp(1.39)} = 0.06$$

$$y_1^0 = \frac{\exp(1.39)}{\exp(-1.39) + \exp(1.39)} = 0.94$$

The estimated y_t^1 and y_t^0 for $t = 1,\ldots, 20$ are shown in Table 4.3.

4.2 Artificial Neural Network

4.2.1 Simplest Network

In Fig. 4.8, a very simple neural network is shown with only one input and one output variables and one hidden node. Note that no bias term is included for simplicity. This can be expressed as

$$Y = \alpha_1 h(s) = \alpha_1 [w_1 x] \tag{4.20}$$

where $s = w_1 x$ and the linear activation function is employed for $h(s)$. Now, the parameters to estimate are α_1 and w_1.

$$Y = \alpha_1 h(s) = \alpha_1 \left[\{1 + \exp(-w_1 x)\}^{-1} \right] \tag{4.21}$$

where the sigmoid activation function is employed for h(s).

By including the bias term, as shown in Fig. 4.9, it can be expressed as for the one with the linear activation function:

$$Y = \alpha_1 h(s) + \alpha_0 = \alpha_1 [w_1 x + w_0] + \alpha_0 \tag{4.22}$$

and for the one with the sigmoid activation function:

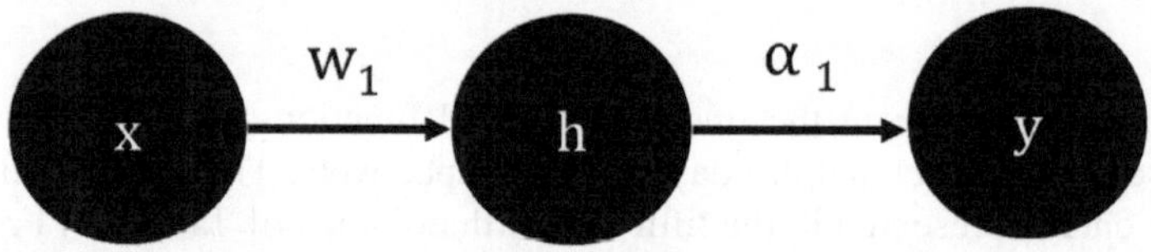

Fig. 4.8 Simplest neural network with one input variable, one hidden node, and one output variable. Note that the current network is the simplest network even without bias term

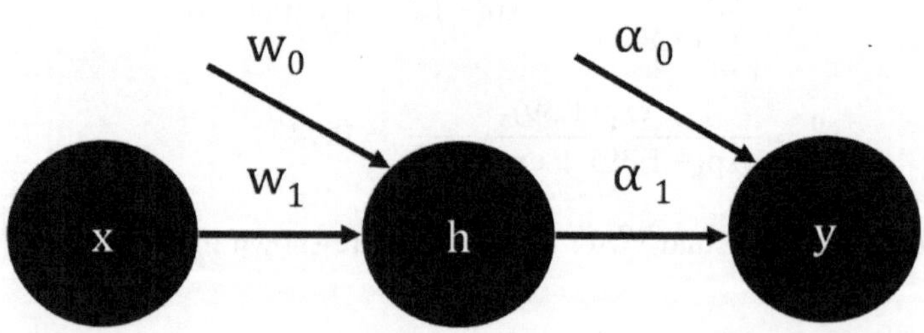

Fig. 4.9 Simplest neural network with one input variable, one hidden node, and one output variable as well as bias terms

$$Y = \alpha_1 h(s) + \alpha_0 = \alpha_1\left[\{1 + \exp(-w_1 x - w_0)\}^{-1}\right] + \alpha_0 \tag{4.23}$$

Example 4.4 Estimate the output layer value for the simplest network in Fig. 4.8. Here, w_1 and α_1 is 0.5 and 0.2, respectively and x = 1.3.

Solution: $y = \alpha_1\left[\{1 + \exp(-0.5 \times 1.3)\}^{-1}\right] = 0.2 \times 0.657 = 0.131.$

Example 4.5 Estimate the output layer value for the simplest network in Eq. (4.22) with the dataset. Here, $w_0$and α_0 are 0.3 and 0.1 and the other values are the same as in Example 4.4.

Solution:

$$\hat{y} = 0.2 \times \left[\{1 + \exp(-0.5 \times 1.3 - 0.3)\}^{-1}\right] + 0.1 = 0.244$$

As one can see from Examples 4.4 and 4.5 that the output layer value can be easily defined. Here, it is obvious that the parameters for the network must be estimated and this parameter estimation procedure is called network training or learning.

Neural network learning is also referred to as "connection learning" or the parameters such as w_1 w_1 and α_1 are also called "weights." The connections between nodes are trained and assigned through training. The closer nodes are, the larger the weights are.

To train the network, the feedforward and backward network process must be explained.

4.2.2 *Feedforward and Backward Propagation*

A feedforward neural network is the one that connects the nodes only forward, not forming a cycle as recurrence. In a feedforward network, the information moves in only one forward direction from input to hidden layer and hidden layer to output.

The parameters (or weights) of a neural network can be estimated with a certain predefined criterion, such as minimizing a mean square error between the predicted value and observed output. One of the popular learning methods for estimating parameters is backward propagation (BP), which searches for a minimum of error function by applying the gradient descent method, and iteratively tunes the weights based on the error derivative with respect to each weight.

4.2.3 Network with Multiple Input and Output Variables

Consider an output vector with one dimension $\mathbf{Y} = [\mathbf{Y}_1, \mathbf{Y}_2, \ldots, \mathbf{Y}_N]^T$ with $\mathbf{Y}_i = [Y_i^1, Y_i^2, \ldots, Y_i^{nO}]$ and an input vector, or called a feature matrix, $\mathbf{x} = [\mathbf{x}_1, \mathbf{x}_2, \ldots, \mathbf{x}_N]^T$ in which $\mathbf{x}_i = [x_i^1, x_i^2, \ldots, x_i^{nI}]$ is a vector with *nI* features. Note that *nO* is the number of output variables and *N* is the record length. A simple example of a neural network is presented in Fig. 4.1. Here, $nO = 2$ and $d = 2$ with three hidden nodes as $nH = 3$.

The predicted output $\mathbf{y}_i$ can be expressed as

$$\mathbf{y}_i = g_{HO}(\mathbf{W}_{HO}\mathbf{h}_i + b_{HO}) \tag{4.24}$$

$$\mathbf{h}_i = g_{IH}(\mathbf{W}_{IH}\mathbf{x}_i + b_{IH}) \tag{4.25}$$

where g_{HO} and g_{IH} are the activation functions from hidden to output and from input to hidden layers, respectively. Also, $\mathbf{W}_{HO}$ and $\mathbf{W}_{IH}$ denote the weight parameter matrices and b_{HO} and b_{IH} denote their bias, as shown in Eq. (4.24).

This can be explicitly described as:

$$\begin{bmatrix} y_i^1 \\ y_i^2 \\ \vdots \\ y_i^{nO} \end{bmatrix} = g_{HO}\left\{ \begin{bmatrix} w_{0,1}^{HO} & \cdots & w_{nH,1}^{HO} \\ \vdots & \ddots & \vdots \\ w_{0,nO}^{HO} & \cdots & w_{nH,nO}^{HO} \end{bmatrix} \begin{bmatrix} 1 \\ h_i^1 \\ h_i^2 \\ \vdots \\ h_i^{nH} \end{bmatrix} \right\} \tag{4.26}$$

$$\begin{bmatrix} h_i^1 \\ h_i^2 \\ \vdots \\ h_i^{nH} \end{bmatrix} = g_{IH}\left\{ \begin{bmatrix} w_{0,1}^{IH} & \cdots & w_{d,1}^{IH} \\ \vdots & \ddots & \vdots \\ w_{0,nH}^{IH} & \cdots & w_{d,nH}^{IH} \end{bmatrix} \begin{bmatrix} 1 \\ x_i^1 \\ x_i^2 \\ \vdots \\ x_i^{nI} \end{bmatrix} \right\} \tag{4.27}$$

Note that $w_{0,j}^{HO}$ and $w_{0,k}^{IH}$ are the bias terms for $j = 1,\ldots, nO$ and $k = 1,\ldots, nH$ as b_{HO} and b_{IH}, while others are the weights for the connection between nodes. The network in Fig. 4.1 can be explicitly represented as:

$$\left[y_i^1 \right] = g_{HO}\left\{ \left[\alpha_0 \; \alpha_1 \; \alpha_2 \; \alpha_3 \right] \begin{bmatrix} 1 \\ h_i^1 \\ h_i^2 \\ h_i^3 \end{bmatrix} \right\} \tag{4.28}$$

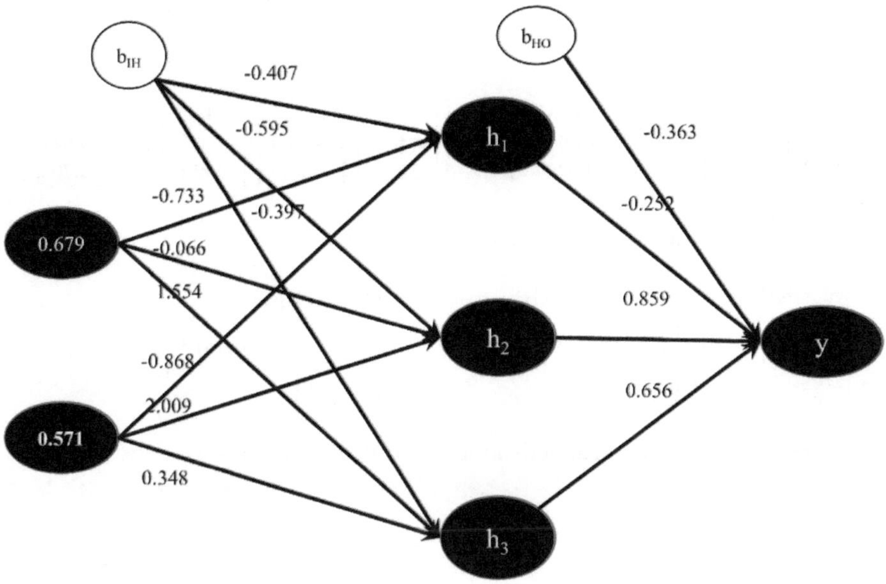

Fig. 4.10 The neural network of Fig. 4.1 with defined weights and biases

$$\begin{bmatrix} h_i^1 \\ h_i^2 \\ h_i^3 \end{bmatrix} = g_{IH}\left\{\begin{bmatrix} w_{0,1} & w_{1,1} & w_{2,1} \\ w_{0,2} & w_{1,2} & w_{2,2} \\ w_{0,3} & w_{1,3} & w_{2,3} \end{bmatrix}\begin{bmatrix} 1 \\ x_i^1 \\ x_i^2 \end{bmatrix}\right\} \tag{4.29}$$

Example 4.6 Estimate the hidden node values and the output values with the biases and weights as shown in Fig. 4.10. Note that the sigmoid function is applied for the activation function of the hidden node and the linear function is used for the output node and the input values are $[0.679, 0.571]^T$. Also, the parameter sets are.

$$W_{IH} = \begin{bmatrix} -0.407 & -0.733 & -0.868 \\ -0.595 & -0.066 & 2.009 \\ -0.397 & 1.554 & 0.348 \end{bmatrix}$$

$$W_{HO} = \begin{bmatrix} -0.363 & -0.252 & 0.859 & 0.656 \end{bmatrix}$$

Solution:

$$\begin{bmatrix} h_i^1 \\ h_i^2 \\ h_i^3 \end{bmatrix} = g_{IH}\left\{\begin{bmatrix} -0.407 & -0.733 & -0.868 \\ -0.595 & -0.066 & 2.009 \\ -0.397 & 1.554 & 0.348 \end{bmatrix}\begin{bmatrix} 1 \\ 0.679 \\ 0.571 \end{bmatrix}\right\} = g_{IH}\left\{\begin{bmatrix} -1.40 \\ 0.51 \\ 0.858 \end{bmatrix}\right\} = \begin{bmatrix} 0.20 \\ 0.62 \\ 0.70 \end{bmatrix}$$

Here,G_{1H} is the sigmoid activation function as $G_{IH}(z) = \frac{1}{1+\exp(-z)}$

$$\begin{bmatrix} y_i^1 \end{bmatrix} = g_{HO}\left\{\begin{bmatrix} -0.363 & -0.252 & 0.859 & 0.656 \end{bmatrix}\begin{bmatrix} 1 \\ 0.2 \\ 0.62 \\ 0.7 \end{bmatrix}\right\} = g_{HO}\{0.583\} = 0.583$$

Here, G_{HO} is the linear activation function as $g_{HO}(z) = z$.

4.2.4 *Python Coding of the Simple Network*

At first, the sigmoid function is defined as.xs

```
def Sigmoid(X):
  return 1/(1+np.exp(-X))
```

Note that np is the predefined call function from numpy. The output value of Example 4.4 is estimated as.

```
w1=0.5
a1=0.2
x=1.3

s=w1*x
hs= Sigmoid(s)
y=hs*a1
```

Also, Example 4.5 is worked out as.

```
w0=-0.3

a0=0.1


s=w1*x+w0

hs=Sigmoid(s)

y=hs*a1+a0
```

Example 4.6 with the vector form is calculated as.

```
x=[1,0.679,0.571]

xt=np.reshape(x,(3,1))


Wih=np.reshape([-0.407,-0.733,-0.868,-0.595,-0.066,2.009,-0.397,1.554,0.348],(3,3))

Who=[-0.363,-0.252,0.859,0.656]


hp=np.dot(Wih,xt)


H=Sigmoid(hp)

H2=np.zeros((4,1))

H2[0]=1.0

H2[1:4]=H

y=np.dot(Who,H2)
```

Overall the program is shown as follows.

[Python 4.1] Program the propagation procedure in Example 4.4, Example 4.5, and Example 4.6

```
import numpy as np
import pandas as pd

def Sigmoid(X):
   return 1/(1+np.exp(-X))

# Example 4.2
w1=0.5
a1=0.2
x=1.3

s=w1*x
hs= Sigmoid(s)
y=hs*a1
```

```
# Example 4.3
w0=-0.3
a0=0.1

s=w1*x+w0
hs=Sigmoid(s)
y=hs*a1+a0

# Example 4.4

x=[1,0.679,0.571]
xt=np.reshape(x,(3,1))

Wih=np.reshape([-0.407,-0.733,-0.868,-0.595,-0.066,2.009,-0.397,1.554,0.348],(3,3))
Who=[-0.363,-0.252,0.859,0.656]

hp=np.dot(Wih,xt)

H=Sigmoid(hp)
H2=np.zeros((4,1))
H2[0]=1.0
H2[1:4]=H
y=np.dot(Who,H2)
```

4.3 Exercises

Estimate CEE and MSE for the y_p data in Table 4.1 but employing the observation as

1	0	1
0	1	0
0	0	1

Create a one-hot encoding matrix for the temperature data in the third column of Table 4.2 by defining hot, cold, and normal as $\geq$ 19, $\leq$ 18, and others, respectively.

Perform the softmax as in Example 4.3 but with the following parameters as $\mathbf{w}$ = [-0.1, 1, 0.5]^T and $\mathbf{b}$ = [0.5,-1, 0.5]^T

Estimate the hidden node value and the output values with the same dataset and parameters in Example 4.6 but employing g_{IH} as tanh and g_{HO} as tanh as well.

Chapter 5
Training a Neural Network

Abstract A neural network is composed of a number of parameters, also called weights. The neural network training is the estimation procedure of those parameters. In this chapter, the training procedure of a neural network is described based on the gradient descent method and the backpropagation algorithm.

Training a network is the procedure to estimate all parameters with input and output datasets, also called network learning. Mainly, there are two approaches to training, supervised and unsupervised learning. An unsupervised algorithm is to train the network with only the input data set and is usually employed to investigate the characteristics of data or preprocess the data.

In contrast, the supervised algorithm uses input and output pair datasets. There are several supervised learning algorithms, such as backpropagation, radial basis function, conjugate gradient, cascade correlation, etc. In this book, only supervised learning with a backpropagation algorithm is used due to its appropriateness to deep learning. The supervised algorithm of the neural network generally proceeds as the following steps.

A. Initialize the weights of the network.
B. Obtain the input and output data from the training dataset and estimate the output from the network and calculate the errors and a selected loss function discussed in Chap. 4.
C. Update the weights by minimizing the selected loss function.
D. Repeat steps B and C until a stopping criterion is satisfied.

5.1 Initialization

As mentioned, the weights of a network must be initialized with adequate values for training as the step of A. Since the input and output datasets are standardized, the weights must also be compressed usually from −3 to + 3. The rule of thumb is to set the range between −3/nI and + 3/nI, that is,

T. Lee et al., *Deep Learning for Hydrometeorology and Environmental Science*, Water Science and Technology Library 99,
https://doi.org/10.1007/978-3-030-64777-3_5

$$w^0 \sim \mathrm{Unif}(-\frac{3}{\mathrm{nI}}, +\frac{3}{\mathrm{nI}}) \tag{5.1}$$

where Unif(a, b) indicates the uniform random number between a and b, nI is the number of input variables. If not restricted, network training may get slower.

The initialization of weights can be done by generating a random number. See the example.

Example 5.1 Initialize the simplest network in Eq. (4.21).

Solution:

The simplest network has two parameters α_1 and w_1. It can be simply simulated from a uniform number as

$$w^1 \sim \mathrm{Unif}\left(-\frac{3}{1}, +\frac{3}{1}\right)$$

$$\alpha_1 \sim \mathrm{Unif}\left(-\frac{3}{1}, +\frac{3}{1}\right)$$

Assume the simulated weights are 2.3 and -1.5 for α_1 and w_1, respectively. Then

$Y = 2.3h(s) = 2.3\left[\{1 + \exp(-1.5x)\}^{-1}\right]$

This can be programmed with python as:

[Python 5.1] Initialize the weights for the 9 values of the network in Figure 4.1 with the range -3 and 3.

```
import numpy as np

nI=2

num=9

w_init=np.random.rand(num)*(3/nI*2)-(3/nI)
```

Note that w_init values are in the range (−3/nI, + 3/nI), that is, (−1.5, + 1.5) with nI = 2.

5.2 Gradient Descent

In training a network, the weights must be iteratively updated, starting from the initialized weights to find the optimum value of an objective function. Here a loss function (see Chap. 4) should be used, which can be denoted as

$$w^{t+1} = w^t - \eta \frac{\partial L(w)}{\partial w} \tag{5.2}$$

where η is the learning rate that controls how quickly a network reaches a solution and $L(.)$ is a loss function, one should choose to train a network for the weight (w).

As the iteration number (t) becomes sufficiently large, the target weight w reaches the minimum value of the target loss function $L(w)$. It is because the minimum is located at the lowest place of the loss function curve with weight w and the weight moves to the minimum as much as $\eta \frac{\partial L(w)}{\partial w}$. The direction to move can be calculated with the gradient of the loss function, that is, the derivative of the loss function with the weight and how much it moves is found with the learning rate. Note that the subtraction is used in Eq. (5.2) because the derivative (i.e., $\frac{\partial L(w)}{\partial w}$) is the gradient indicating how much a function increases along with a variable (here, w).

With a large learning rate (η), a network learns faster but more highly changes to be on a sub-optimum and vice versa. To illustrate the effect of different learning rates, Fig. 5.2 shows how fast its training reaches the optimum position with $\eta = 0.1, 0.01,$ and 0.001. Note that with a high learning rate as $\eta = 0.1$, the training does not converge to the optimum.

Two possible approaches can be applied to configure the learning rates as performing sensitivity analysis with diagnostic plots of gridded learning rates and employing momentum or adaptive learning rates. A number of approaches have been studied for the latter and are discussed in the next chapter.

Multiple weight parameters of a network can be updated simultaneously as

$$\mathbf{w}_{t+1} = \mathbf{w}_t - \eta \frac{\partial E}{\partial \mathbf{w}} \tag{5.3}$$

where $\mathbf{w}_t$ is the t th parameter matrix and $\mathbf{w}_0$ must be initialized in advance as discussed. $\frac{\partial E}{\partial \mathbf{w}}$ is the gradient matrix of all the weights and biases.

Example 5.2 Update the weights of the network (i.e. w_1 and α_1) in Fig. 5.1 with the gradient descent method (set μ as 0.9). Here, the current w_1 and α_1 are 0.8 and -0.3, respectively, and the learning rate is 0.01.

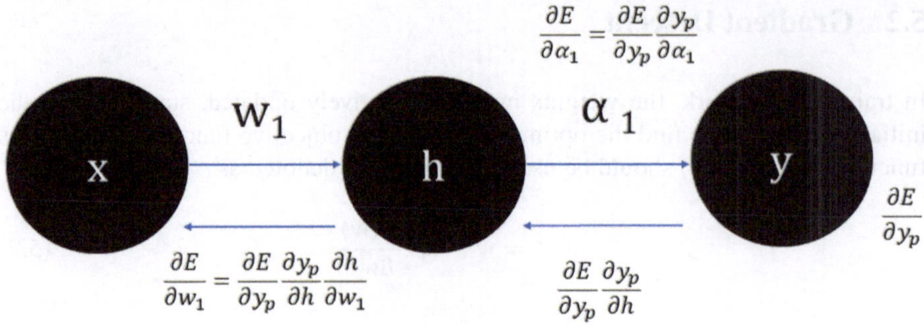

Fig. 5.1 Simplest neural network in Fig. 4.8 with backward gradients

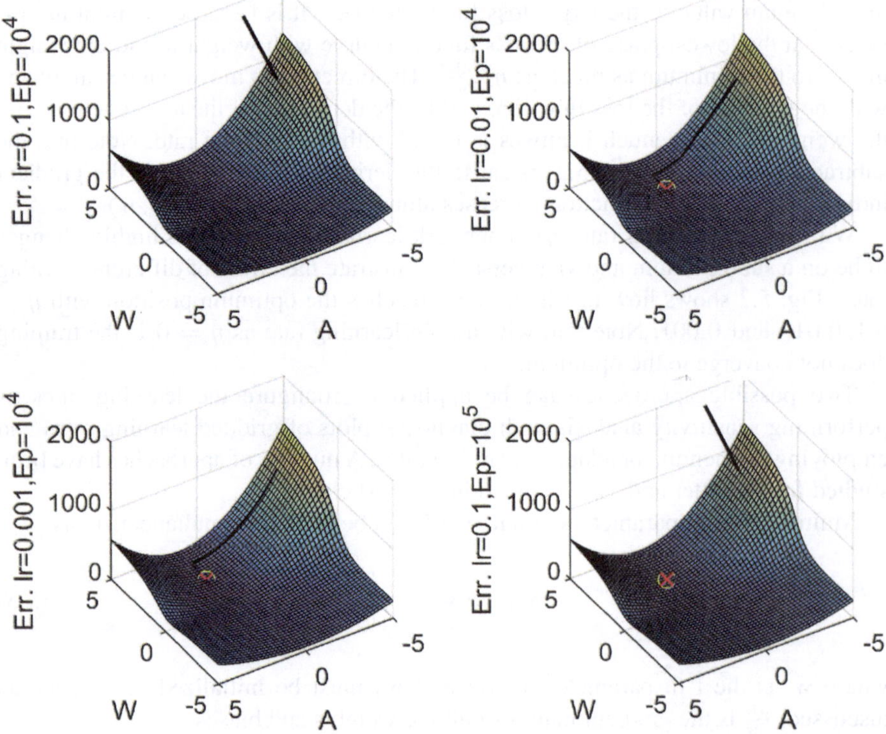

Fig. 5.2 Parameter update with gradient descent with different learning rates (0.1, 0.01, 0.001) and epochs (10^4 and 10^5). Note that when the learning rate is too large as 0.1, the parameters diverge out

Solution:

$\mathbf{w}_0 = \begin{bmatrix} 0.8 \\ -0.3 \end{bmatrix}$

Assume $\frac{\partial E}{\partial w_1} = 0.1$ and $\frac{\partial E}{\partial \alpha_1} = -2$

w_1 should be the same as in Example 5.3

$\mathbf{w}_1 = \mathbf{w}_0 - \eta \frac{\partial E}{\partial \mathbf{w}} = \begin{bmatrix} 0.8 \\ -0.3 \end{bmatrix} - 0.01 \begin{bmatrix} 0.1 \\ -2 \end{bmatrix} = \begin{bmatrix} 0.799 \\ -0.280 \end{bmatrix}$

5.3 Backpropagation

The weights (including biases) of a network are estimated through network learning from a supervised algorithm as the gradient descent in Eqs. (5.2) and (5.3). From these equations, the gradients must be calculated at each step. One of the major methods for calculating the gradients is backpropagation (Rumelhart et al. 1986). The gradient (represented with the partial derivative) for a certain weight is estimated with the chain rule backwardly propagated from the output. The chain rule efficiently works since the neural network is continuously connected like a chain from an input layer to the output layer.

5.3.1 *Simple Network*

For example, the gradient of w_1 in Fig. 5.1 for the error (i.e., $\frac{\partial E}{\partial w_1}$) with a chain rule is expressed as

$$\frac{\partial E}{\partial w_1} = \frac{\partial E}{\partial y_p} \frac{\partial y_p}{\partial h} \frac{\partial h}{\partial w_1} \tag{5.4}$$

The chain rule follows the connection of the network in Fig. 5.1. The error is expressed as a function of the predicted variable ($y_{p,i}$) as

$$E = \frac{1}{2} \sum_{i=1}^{N} \left(y_{o,i} - y_{p,i} \right)^2 = \frac{1}{2} \sum_{i=1}^{N} e_i^2 \tag{5.5}$$

Note that the L2 loss function in Eq. (4.12) is used by multiplying ½ (divided out in derivative) for the error in this case. The predicted variable ($y_{p,i}$) is a function of the hidden node as

$$y_{p,i} = \alpha_1 h(s_i) = \alpha_1 h(w_1 x_i) \tag{5.6}$$

where $s_i = w_1 x_i$. Assume the sigmoid function in Eq. (4.3) is used as an activation function and its derivative (i.e., $h(1\text{-}h)$). This can be explicitly expressed as

$$\frac{\partial E}{\partial w_1} = \frac{\partial E}{\partial y_p}\frac{\partial y_p}{\partial h}\frac{\partial h}{\partial w_1} = -\sum\nolimits_{i=1}^{N} e_i \alpha_1 h(1-h)x_i \tag{5.7}$$

where

$$\frac{\partial E}{\partial y_p} = -\sum\nolimits_{i=1}^{N} e_i = -\sum\nolimits_{i=1}^{N} y_{o,i} - y_{p,i} \tag{5.8}$$

$$\frac{\partial y_p}{\partial h} = \frac{\partial(\alpha_1 h)}{\partial h} = \alpha_1 \tag{5.9}$$

$$\frac{\partial h}{\partial w_1} = \frac{\partial h}{\partial s}\frac{\partial s}{\partial w_1} = h(1-h)\frac{\partial s}{\partial w_1} = h(1-h)x_i \tag{5.10}$$

since $\frac{\partial h}{\partial s} = h(1-h)$ from Eq. (4.4).

Also, the gradient of α_1 in Fig. 5.1 for the error (i.e., $\frac{\partial E}{\partial \alpha_1}$) is expressed as:

$$\frac{\partial E}{\partial \alpha_1} = \frac{\partial E}{\partial y_p}\frac{\partial y_p}{\partial \alpha_1} = -\sum\nolimits_{i=1}^{N} e_i h(w_1 x_i) = -\sum\nolimits_{i=1}^{N} (y_{o,i} - y_{p,i})h(w_1 x_i) \tag{5.11}$$

where

$$\frac{\partial y_p}{\partial \alpha_1} = h(s_i) = h(w_1 x_i) \tag{5.12}$$

Example 5.3 Update the weights of the network (i.e. w_1 and α_1) in Fig. 5.1. Here, the current w_1 and α_1 are 0.8 and -0.3, respectively, and the learning rate is 0.01. Note that the PDO dataset was employed with minmax scaling staring in January 1900 in Table 5.1. Let $Y_\mathrm{i} = Y_\mathrm{i}$ and $X_\mathrm{i} = Y_{i\text{-}1}$ where, Y_i is the ith PDO value.

Solution:

The input (*xi*) and output data are shown in the second and third columns of Table 5.1.

Table 5.1 Example of backpropagation with the simple network in Fig. 5.1. The monthly PDO dataset was employed with minmax scaling, staring in January 1900

Weights	w_1^{t-1}	α_1^{t-1}	η	w_1^t	α_1^t		
	0.8	−0.3	0.01	0.799	−0.280		
No	xi	yo,i	yp,i	$ei = yo,i - yp,i$	$h(si)$	$ei\ h(si)$	$\alpha_1^{t-1} * eh(1-h) * x$
1	0.18	0.80	−0.16	0.96	0.54	0.515	−0.013
2	0.80	0.34	−0.20	0.54	0.65	0.351	−0.029
3	0.34	0.20	−0.17	0.37	0.57	0.210	−0.009
4	0.20	0.19	−0.16	0.35	0.54	0.190	−0.005
5	0.19	−0.19	−0.16	−0.03	0.54	−0.015	0.000
6	−0.19	−0.58	−0.14	−0.44	0.46	−0.204	−0.006
7	−0.58	−0.78	−0.12	−0.66	0.39	−0.256	−0.027
8	−0.78	0.58	−0.10	0.68	0.35	0.239	0.036
9	0.58	0.39	−0.18	0.57	0.61	0.353	−0.024
10	0.39	0.84	−0.17	1.01	0.58	0.585	−0.029
					−Sum	−1.967	0.106

$$\frac{\partial E}{\partial w_1} = -\sum_{i=1}^{N} e_i \alpha_1 h(1-h)x_i = -\{0.96(-0.3)0.54(1-0.54)0.18 + \cdots\} = 0.106$$

$$w_1^t = w_1^{t-1} - \eta \frac{\partial L(w_1)}{\partial w_1} = 0.8 - 0.01(0.106) = 0.799$$

$$\frac{\partial E}{\partial \alpha_1} = -\sum_{i=1}^{N} e_i h = -\{0.96 \times 0.54 + \cdots\} = -1.967$$

$$\alpha_1^t = \alpha_1^{t-1} - \eta \frac{\partial L(\alpha_1)}{\partial \alpha_1} = -0.3 - 0.01(-1.967) = -0.280$$

In Fig. 5.2, the total path of the gradient descent method with 100 data values of the PDO starting in January 1900 is presented with different learning rates and the number of epochs with the initial point as $w_1 = 4$ and $\alpha_1 = -4$. As shown in the top left panel and the bottom right panel, the learning rate 0.1 is too large to have appropriate learning. Due to the large learning rate, the learning is diverged. With the smaller learning rate as 0.001, the learning rate can reach the optimum value as shown in the bottom left panel. The gradient descent learning presents that the variable α_1 has a much highly updated value followed by w_1 when α_1 is reached at the valley (see the top right panel).

5.3.2 *Full Neural Network*

The backpropagation procedure for the full neural network example in Fig. 5.3 is shown here. Assume that the activation functions of the hidden node and output nodes are sigmoid and linear function, respectively. The forward network can be expressed as

$$y = \alpha_0 + \alpha_1 h_1(w_{01} + w_{11}x_1 + w_{21}x_2) + \alpha_2 h_2(w_{02} + w_{12}x_1 + w_{22}x_2) + \alpha_3 h_3(w_{03} + w_{13}x_1 + w_{23}x_2) \tag{5.13}$$

The gradients of weights are described as follows. For the bias and weights between input and hidden layers are:

$$\frac{\partial E}{\partial w_{0j}} = \frac{\partial E}{\partial y_p}\frac{\partial y_p}{\partial h_j}\frac{\partial h_j}{\partial w_{0j}} = -\sum\nolimits_{i=1}^{N} e_i \alpha_j h_j(1-h_j) \tag{5.14}$$

where

$$\frac{\partial E_i}{\partial y_p} = -\left(y_{o,i} - y_{p,i}\right) = -e_i \tag{5.15}$$

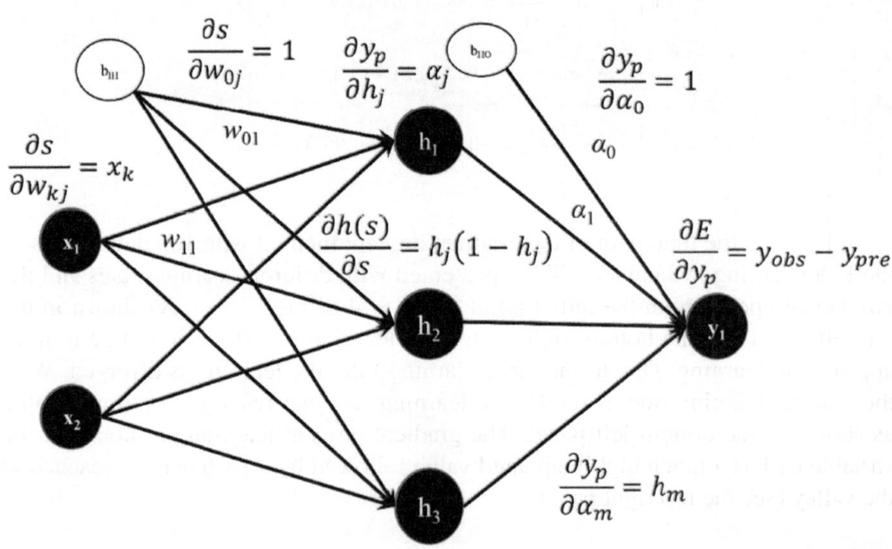

Fig. 5.3 Example of neural network in Fig. 4.1 with backward gradients. Note that the gradient of each weight and bias can be calculated with backward propagation. For example, the gradient $\frac{\partial E_i}{\partial w_{kj}} = \frac{\partial E_i}{\partial y_p}\frac{\partial y_p}{\partial h_j}\frac{\partial h_j}{\partial w_{kj}} = \left(y_{o,i} - y_{p,i}\right)\alpha_j h_j(1-h_j)x_k$ is the multiplication of the derivatives of the feedforward network

$$\frac{\partial y_p}{\partial h_j} = \alpha_j \tag{5.16}$$

and

$$\frac{\partial h}{\partial w_{0j}} = \frac{\partial h(s)}{\partial s}\frac{\partial s}{\partial w_{0j}} = h_j(1 - h_j) \tag{5.17}$$

since $s = w_{0j} + w_{1j}x_1 + w_{2j}x_2$ and $\frac{\partial s}{\partial w_{0j}} = 1$. Note that $j = 1, 2, 3$ (i.e., $w_{01}, w_{02},$ and w_{03}).

$$\frac{\partial E}{\partial w_{kj}} = \frac{\partial E}{\partial y_p}\frac{\partial y_p}{\partial h_j}\frac{\partial h_j}{\partial w_{kj}} = -\sum_{i=1}^{N} e_i\alpha_j h_j(1 - h_j)x_k \tag{5.18}$$

for $k = 1$ and 2. For example, $k = 2$ and $j = 3$

$$\frac{\partial E}{\partial w_{23}} = \frac{\partial E}{\partial y_p}\frac{\partial y_p}{\partial h_3}\frac{\partial h_3}{\partial w_{23}} = -\sum_{i=1}^{N} e_i\alpha_3 h_3(1 - h_3)x_2 \tag{5.19}$$

Also, for the weights and bias between hidden and output layers

$$\frac{\partial E}{\partial \alpha_0} = \frac{\partial E}{\partial y_p}\frac{\partial y_p}{\partial \alpha_0} = -\sum_{i=1}^{N} e_i \tag{5.20}$$

where $\frac{\partial y_p}{\partial \alpha_0} = 1$.

Also,

$$\frac{\partial E}{\partial \alpha_m} = \frac{\partial E}{\partial y_p}\frac{\partial y_p}{\partial \alpha_m} = -\sum_{i=1}^{N} e_i h_m \tag{5.21}$$

since $\frac{\partial y_p}{\partial \alpha_m} = h_m$ and $m = 1, 2,$ and 3.

Example 5.4 Update the weights of the network in Fig. 5.3. The network consists with two input nodes, one output node and three hidden nodes. The observed input (X) and output (Y) dataset are shown in Table 5.2 in the 2nd to 4th columns. The initial weights are shown in the second column of Table 5.3. Note that the initial values are randomly generated from Eq. (5.1). The predicted values with the initial set are shown in the 5th column of Table 5.2 as well as the error in the 6th column. The learning rate is 0.01.

Solution:

The weights for the hidden node to the output node (i.e. α_0 and α_{m}) can be estimated as Eqs. (5.20) and (5.21).

$$\frac{\partial E}{\partial \alpha_0} = -\sum_{i=1}^{N} e_i = 4.66$$

The negative sum of the error is shown at the bottom of the last column of Table 5.2 as 4.66.

$$\frac{\partial E}{\partial \alpha_1} = -\sum_{i=1}^{N} e_i h_m = 1.899$$

Note that the negative sum of $e_i h_m$ is shown at the bottom of the left side in Table 5.5 and the estimated $h_m(s_i)$ are shown at the right side of Table 5.4 as well as s_i at the left side.The other gradients of $\frac{\partial E}{\partial \alpha_2}$ and $\frac{\partial E}{\partial \alpha_3}$ are also shown at the bottom of the left side in Table 5.5 as 1.124 and 2.988, respectively.

The weights for the input node to the hidden node (i.e. w_{0j} and w_{kj}) can be estimated as Eqs. (5.17) and (5.18).

$$\frac{\partial E}{\partial w_{01}} = -\sum_{i=1}^{N} e_i \alpha_1 h_1 (1 - h_1) = -1.126$$

The negative sum of $e_i \alpha_1 h_1 (1 - h_1)$ is shown at the bottom of the right side in Table 5.5. Also, $\frac{\partial E}{\partial w_{02}}$ and $\frac{\partial E}{\partial w_{03}}$ are −0.272 and 1.476.

Solution:

$$\frac{\partial E}{\partial w_{11}} = -\sum_{i=1}^{N} e_i \alpha_1 h_1 (1 - h_1) x_1 = -0.639$$

The detailed estimation is presented in the 2nd column of Table 5.6 and its negative sum is presented at the bottom. $\frac{\partial E}{\partial w_{12}}$ and $\frac{\partial E}{\partial w_{13}}$ are −0.155 and 0.838, respectively.

$$\frac{\partial E}{\partial w_{21}} = -\sum_{i=1}^{N} e_i \alpha_1 h_1 (1 - h_1) x_2 = -0.636$$

The detailed estimation is presented in the 5th column of Table 5.6 and its negative sum is presented at the bottom. $\frac{\partial E}{\partial w_{22}}$ and $\frac{\partial E}{\partial w_{23}}$ are −0.153 and 0.835, respectively.

All the parameters are updated with Eq. (5.3), that is,

$$\mathbf{w}_{t+1} = \mathbf{w}_t - \eta \frac{\partial E}{\partial \mathbf{w}}$$

$$\begin{bmatrix} w_{01} \\ w_{11} \\ w_{21} \\ w_{02} \\ w_{12} \\ w_{22} \\ w_{03} \\ w_{13} \\ w_{23} \\ \alpha_0 \\ \alpha_1 \\ \alpha_2 \\ \alpha_3 \end{bmatrix}_t = \begin{bmatrix} w_{01} \\ w_{11} \\ w_{21} \\ w_{02} \\ w_{12} \\ w_{22} \\ w_{03} \\ w_{13} \\ w_{23} \\ \alpha_0 \\ \alpha_1 \\ \alpha_2 \\ \alpha_3 \end{bmatrix}_{t-1} - \eta \begin{bmatrix} \frac{\partial E}{\partial w_{01}} \\ \frac{\partial E}{\partial w_{11}} \\ \vdots \\ \frac{\partial E}{\partial \alpha_2} \\ \frac{\partial E}{\partial \alpha_3} \end{bmatrix} = \begin{bmatrix} -0.76 \\ 1.18 \\ -0.51 \\ -1.03 \\ 0.86 \\ -1.07 \\ 1.18 \\ -0.82 \\ -0.25 \\ 0.63 \\ -1.00 \\ -0.32 \\ 1.38 \end{bmatrix} - 0.01 \begin{bmatrix} 1.13 \\ 0.64 \\ 0.64 \\ 0.27 \\ 0.15 \\ 0.15 \\ -1.48 \\ -0.84 \\ -0.84 \\ -4.66 \\ -1.90 \\ -1.12 \\ -2.99 \end{bmatrix} = \begin{bmatrix} -0.75 \\ 1.19 \\ -0.50 \\ -1.03 \\ 0.87 \\ -1.07 \\ 1.17 \\ -0.83 \\ -0.26 \\ 0.58 \\ -1.02 \\ -0.33 \\ 1.35 \end{bmatrix}$$

The overall result is shown in Fig. 5.4 indicating that training works well and shows reasonable performance.

Table 5.2 Dataset of observed X, Y, and predicted Y as well as their errors for the example of backpropagation with the neural network in Fig. 5.3

No	X_1	X_2	$y_{o,i}$	$y_{p,i}$	$e_i = y_{o,i} - y_{p,i}$
1	0.51	0.69	0.58	1.07	−0.50
2	0.69	0.58	0.56	0.95	−0.40
3	0.58	0.56	0.61	1.02	−0.41
4	0.56	0.61	0.60	1.04	−0.44
5	0.61	0.60	0.64	1.00	−0.36
6	0.60	0.64	0.53	1.02	−0.49
7	0.64	0.53	0.47	0.98	−0.51
8	0.53	0.47	0.54	1.04	−0.50
9	0.47	0.54	0.44	1.08	−0.64
10	0.54	0.44	0.62	1.03	−0.41
				−Sum	4.66

Table 5.3 Initial $(t-1)$ and updated (t) weights and biases of the neural network in Fig. 5.3

Weights	t−1†	t
w_{01}	−0.759	−0.748
w_{11}	1.183	1.190
w_{21}	−0.506	−0.499
w_{02}	−1.034	−1.031
w_{12}	0.865	0.866
w_{22}	−1.069	−1.068
w_{03}	1.183	1.169
w_{13}	−0.818	−0.826
w_{23}	−0.247	−0.255
α_0	0.630	0.583
α_1	−1.002	−1.021
α_2	−0.320	−0.331
α_3	1.378	1.348

†This initial weights and biases were randomly generated as in Eq. (5.1)

Table 5.4 Dataset of the estimated S and h for the example of backpropagation with the neural network in Fig. 5.3

No	$S_j = w_{0j} + w_{1j}x_1 + w_{2j}x_2$			$h(S_j)$		
	S_1	S_2	S_3	*h 1*	*h 2*	*h3*
1	−0.503	−1.331	0.594	0.377	0.209	0.644
2	−0.231	−1.051	0.475	0.442	0.259	0.617
3	−0.360	−1.130	0.576	0.411	0.244	0.640
4	−0.413	−1.211	0.577	0.398	0.230	0.640
5	−0.334	−1.142	0.533	0.417	0.242	0.630
6	−0.376	−1.201	0.537	0.407	0.231	0.631
7	−0.268	−1.043	0.530	0.433	0.261	0.630
8	−0.376	−1.084	0.637	0.407	0.253	0.654
9	−0.473	−1.201	0.664	0.384	0.231	0.660
10	−0.347	−1.043	0.633	0.414	0.261	0.653
−Sum	**3.680**	**11.438**	**−5.756**	**−4.091**	**−2.420**	**−6.400**

Table 5.5 Dataset of the estimated $\frac{\partial E}{\partial \alpha_m}$ and $\frac{\partial E}{\partial w_{0j}}$ for the example of backpropagation with the neural network in Fig. 5.3

No	$\frac{\partial E}{\partial \alpha_m}$			$\frac{\partial E}{\partial w_{0j}}$		
	$e_i h_1(S_1(i))$	$e_i h_2(S_2(i))$	$e_i h_3(S_3(i))$	$\alpha_1 e_i h_1(1- h1)$	$\alpha_2 e_i h_2(1- h2)$	$\alpha_2 e_i h_2(1- h2)$
1	−0.188	−0.104	−0.321	0.117	0.026	−0.157
2	−0.176	−0.103	−0.245	0.098	0.024	−0.130
3	−0.167	−0.099	−0.261	0.099	0.024	−0.129
4	−0.176	−0.101	−0.283	0.106	0.025	−0.140
5	−0.151	−0.088	−0.229	0.088	0.021	−0.116
6	−0.200	−0.114	−0.310	0.119	0.028	−0.158
7	−0.220	−0.132	−0.319	0.125	0.031	−0.163
8	−0.205	−0.127	−0.329	0.122	0.030	−0.157
9	−0.244	−0.147	−0.420	0.151	0.036	−0.197
10	−0.172	−0.108	−0.270	0.101	0.026	−0.129
−Sum	**1.899**	**1.124**	**2.988**	**−1.126**	**−0.272**	**1.476**

Table 5.6 Dataset of the estimated $\frac{\partial E}{\partial w_{1j}}$ and $\frac{\partial E}{\partial w_{2j}}$ for the example of backpropagation with the neural network in Fig. 5.3

No	$\frac{\partial E}{\partial w_{1j}}$			$\frac{\partial E}{\partial w_{2j}}$		
	$\alpha_1 e_i h_1(1- h1) x_1$	$\alpha_2 e_i h_2(1- h2)\, x_1$	$\alpha_2 e_i h_2(1- h2)\, x_1$	$\alpha_1 e_i h_1(1- h1) x_2$	$\alpha_2 e_i h_2(1- h2)\, x_2$	$\alpha_2 e_i h_2(1- h2)\, x_2$
1	0.060	0.013	−0.080	0.081	0.018	−0.109
2	0.068	0.017	−0.090	0.057	0.014	−0.075
3	0.057	0.014	−0.074	0.055	0.013	−0.072
4	0.059	0.014	−0.078	0.065	0.015	−0.086
5	0.054	0.013	−0.072	0.053	0.013	−0.070
6	0.071	0.017	−0.094	0.076	0.018	−0.101
7	0.080	0.020	−0.104	0.066	0.016	−0.086
8	0.064	0.016	−0.083	0.058	0.014	−0.074
9	0.071	0.017	−0.093	0.081	0.019	0.106
10	0.054	0.014	−0.070	0.045	0.011	−0.057
−Sum	**−0.639**	**−0.155**	**0.838**	**−0.636**	**−0.153**	**0.835**

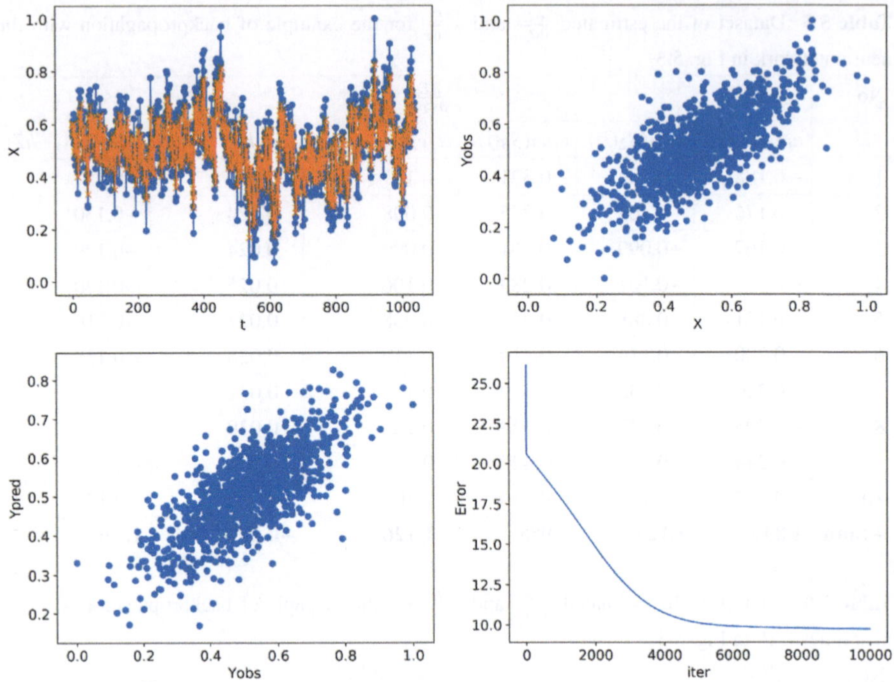

Fig. 5.4 Results of the PDO prediction with the neural network model in Fig. 5.3. The left top panel presents the time series of the observed monthly PDO dataset (blue line) from https://www.ncdc.noaa.gov/teleconnections/pdo/ as well as the predicted values from the neural network example of Example 5.4 as shown in Fig. 5.3. The right top panel shows the scatter plot between the predictand ($x = y_{t-1}$) and the predictand (y_t), while the bottom left panel presents the scatter plot between the observed value of y_t and the predicted value. The right bottom panel shows how the error function varies along with the iteration

5.3.3 *Python Coding of Network*

The following presents the python code in Example 5.4.

```
[Python 5.2] Program the propagation procedure in [Example 5.4]

import numpy as np
import pandas as pd
# Function for Feedforward network to estimate Yprediction
def Sigmoid(X):
  return 1/(1+np.exp(-X))

def dSigmoid(X):
  y=Sigmoid(X)
  return y*(1-y)

def FFNetwork(W,A,Xt,nX,nH): # Feedforward network to estimate Yprediction
  ht=np.zeros((nX,nH))
  for j in range(nH):
      ht[:,j]=Sigmoid(np.dot(W[j,:],Xt.T))
  return np.dot(A[1:],ht.T)+A[0], ht
```

```
# Gradient Estimation
def dEdw0j(er,alp_j,hs_j):
    return -sum(alp_j*er*hs_j*(1-hs_j))
def dEdw_kj(er,alp_j,hs_j,xk):
    return -sum(alp_j*er*hs_j*(1-hs_j)*xk)
def dEda0(er):
    return -sum(er)
def dEdam(er,hm):
    return -sum(er*hm)

# Estimate All the gradients
def gradient_weights(W,A,Xt,Yobs,Ypre,ht,nH):
    nX,nl=np.shape(Xt)
    nl=nl-1

    Err=Yobs-Ypre
    dA=np.zeros(len(A))
    dW=np.zeros(np.shape(W))

    dA[0]=dEda0(Err)
    for m in range(nH):
        dA[m+1]=dEdam(Err,ht[:,m])

    for j in range(nH):
        dW[j,0]=dEdw0j(Err,A[j+1],ht[:,j])
        for k in range(nl):
            dW[j,k+1]=dEdw_kj(Err,A[j+1],ht[:,j],Xt[:,k+1])

    return dA,dW

[Python 5.2 ] Continue

nl=2 # number of input variables
nH=3 # number of hidden nodes
num_para=13
#X_train #input data N*nl    matrix
#Ypre       #output data N*1
# parameter initialization
w_init=np.random.rand(num_para)*(3/nl*2)-(3/nl)
w01,w11,w21,w02,w12,w22,w03,w13,w23,a0,a1,a2,a3=w_init
# reivse input matrix including the one vector for bias
X1=np.ones((len(X_train),1),dtype='float')
X_t=np.concatenate((X1,X_train),axis=1)

nX=len(X_t) # number of data
# Set the weights (input to hidden layers W) to the matrix form
Wn=[w01,w11,w21,w02,w12,w22,w03,w13,w23]
W=np.reshape(Wn,(3,3))
# Set the weights (hidden to output layers A) to the matrix form
A=[a0,a1,a2,a3]

# Repeated procedure for updating the gradients
lr=0.0001#learning rate
count=0
```

```
count=0
Ncount=10000
L2=np.zeros(Ncount)
A0all=np.zeros(Ncount)
Ypre,ht=FFNetwork(W,A,X_t,nX,nH)
Yobs=Y_train
while count<Ncount:
  dA,dW= gradient_weights(W,A,X_t,Yobs,Ypre,ht,nH)
  Ypre,ht=FFNetwork(W,A,X_t,nX,nH)
  L2[count]=np.sum((Yobs-Ypre)**2)
  A=A-lr*dA
  W=W-lr*dWcount+=1

fig=plt.figure(figsize=(15,12))
ax1=fig.add_subplot(2,2,1)
ax2=fig.add_subplot(2,2,2)
ax3=fig.add_subplot(2,2,3)
ax4=fig.add_subplot(2,2,4)
ax1.plot(Yobs,marker='o')
ax1.plot(Ypre,marker='x'), ax1.set_xlabel('t'),ax1.set_ylabel('X')
ax2.scatter(X_t[:,2],Yobs), ax2.set_xlabel('X'),ax2.set_ylabel('Yobs')
ax3.scatter(Yobs,Ypre), ax3.set_xlabel('Yobs'),ax3.set_ylabel('Ypred')
ax4.plot(L2[10:]), ax4.set_xlabel('iter'),ax4.set_ylabel('Error')
```

5.4 Exercises

1. Update the weights of the simplest network in Fig. 5.1 (i.e., α_1 and w_1) for the PDO dataset in Table 5.1. The learning rate is 0.1 and the number of epochs is 5. The initial values of the parameters α_1 and w_1 are 0.5 and -0.5, respectively. Here, $Y_i = y_i$ and $X_i = y_{i-1}$
2. Derive $\frac{\partial E}{\partial w_{11}}$ and $\frac{\partial E}{\partial w_{11}}$ for the simplest network in Fig. 5.1 employing tanh for the hidden activation function.
3. Apply the network in Fig. 5.3 to the data from the second to the fourth columns of Table 5.1 but employing tanh for the hidden activation function. The same data is used for initial weights as the second column of Table 5.3.

Reference

Rumelhart DE, Hinton GE, Williams RJ (1986) Learning representations by back-propagating errors. Nature 323(6088):533–536. https://doi.org/10.1038/323533a0

Chapter 6
Updating Weights

Abstract Training a neural network is to update the weights to minimize a specified loss function and the gradient descent method has been employed. However, the number of weights exponentially grows, especially in a deep learning machine. In recent years, several methods updating weights have been developed to improve the speed of convergence and to find the best trajectory to reach the optimum of the employed loss function for a network. In this chapter, those methods for updating weights are explained.

Training a neural network is to update the weights to minimize a loss function and the gradient descent method has been employed. However, the number of weights exponentially grows, especially in a deep learning machine. In recent years, several methods updating weights have been developed to improve the speed of convergence and to find the best trajectory to reach the optimum of the employed loss function for a network.

6.1 Momentum

The use of momentum has a physical analogy, such as rolling a ball down the inside of a bowl (Qian 1999). The momentum gets higher on a steeper gradient. The momentum method updates the weights according to the gradient of the current step as well as the gradient of the previous time step. The momentum method can be expressed as:

$$\boldsymbol{w}_{t+1} = \boldsymbol{w}_t + \boldsymbol{m}_t \tag{6.1}$$

$$\boldsymbol{m}_t = \mu \boldsymbol{m}_{t-1} - \eta \left(\frac{\partial E}{\partial \boldsymbol{w}} \right) \tag{6.2}$$

T. Lee et al., *Deep Learning for Hydrometeorology and Environmental Science*, Water Science and Technology Library 99,
https://doi.org/10.1007/978-3-030-64777-3_6

where μ is a moment term, typically set as 0.9 or 0.5. See Eq. (5.3) for the original gradient descent method. The momentum equation with $\mu = 1$ is explicitly expressed as:

$$\boldsymbol{w}_{t+1} = \boldsymbol{w}_t + \mu \boldsymbol{m}_{t-1} - \eta\left(\frac{\partial E}{\partial \boldsymbol{w}}\right) = \cdots = \boldsymbol{w}_t - \eta \sum_{j=1}^{t-1}\left(\frac{\partial E}{\partial \boldsymbol{w}}\right)_j \quad (6.3)$$

The equation implies that the updated portion is the summation of the previous gradient.

Compared to the original gradient descent method, the momentum method moves faster towards convergence. Furthermore, the method is more efficient in traveling a ravine shape (i.e., a surface curves more steeply in one dimension than in another). However, the momentum might be too much, especially for the one with higher gradients, since its movement keeps accumulating. The method is also easy to diverge out from the optimum, even with a relatively small learning rate (η).

Example 6.1 Update the weights of the network (i.e., w_1 and α_1) in Fig. 5.1 with the momentum method (set μ as 0.9). Here, the current w_1 and α_1 are 0.8 and −0.3, respectively, and the learning rate is 0.01 and employ the PDO dataset in the 2nd and 3rd columns of Table 6.1. Set μ as 0.9.

Solution:

Note that for $\boldsymbol{m}_1 = 0.9 \times \boldsymbol{m}_0 - \eta\left(\frac{\partial E}{\partial \boldsymbol{w}}\right)$, $\boldsymbol{m}_0$ is set as zero and $\boldsymbol{w} = \begin{bmatrix} w_1 \\ \alpha_1 \end{bmatrix}$.

Then $\frac{\partial E}{\partial w_1} = 0.106$ and $\frac{\partial E}{\partial \alpha_1} = -1.967$ as shown in Table 6.1

Table 6.1 Dataset of the momentum method for the simple network in Fig. 5.1 in Example 6.1. The PDO dataset was employed with minmax scaling, staring in January 1900

No	x_i	$y_{o,i}$	$y_{p,i}$	e_i	h	eh	$\alpha_1^{t-1}\, eh(1-h)x$
1	0.18	0.80	−0.15	0.95	0.54	0.51	−0.01
2	0.80	0.34	−0.18	0.52	0.65	0.34	−0.03
3	0.34	0.20	−0.16	0.36	0.57	0.20	−0.01
4	0.20	0.19	−0.15	0.34	0.54	0.18	0.00
5	0.19	−0.19	−0.15	−0.04	0.54	−0.02	0.00
6	−0.19	−0.58	−0.13	−0.45	0.46	−0.21	−0.01
7	−0.58	−0.78	−0.11	−0.67	0.39	−0.26	−0.03
8	−0.78	0.58	−0.10	0.68	0.35	0.24	0.03
9	0.58	0.39	−0.17	0.56	0.61	0.35	−0.02
10	0.39	0.84	−0.16	1.00	0.58	0.58	−0.03
					-sum	−1.91	0.10

$$\boldsymbol{m}_1 = \mu \boldsymbol{m}_0 - \begin{bmatrix} \frac{\partial E}{\partial w_1} \\ \frac{\partial E}{\partial \alpha_1} \end{bmatrix} = \begin{bmatrix} 0 \\ 0 \end{bmatrix} - 0.01 \begin{bmatrix} 0.106 \\ -1.967 \end{bmatrix} = \begin{bmatrix} -0.0011 \\ 0.0197 \end{bmatrix}$$

$\boldsymbol{w}_1$ should be the same as in Example 5.3

$$\boldsymbol{w}_1 = \begin{bmatrix} 0.8 \\ -0.3 \end{bmatrix} + \begin{bmatrix} -0.0011 \\ 0.0197 \end{bmatrix} = \begin{bmatrix} 0.799 \\ -0.280 \end{bmatrix}$$

For updating the weights as $\boldsymbol{w}_2$, $y_{p,i}$ is calculated with Eq. (5.6) and $e_i = y_{o,i} - y_{p,i}$.

Then, $\frac{\partial E}{\partial w_1} = 0.1$ and $\frac{\partial E}{\partial \alpha_1} = -1.91$

$$\boldsymbol{m}_2 = 0.9 \times \begin{bmatrix} -0.0011 \\ 0.0197 \end{bmatrix} - 0.01 \begin{bmatrix} 0.10 \\ -1.91 \end{bmatrix} = \begin{bmatrix} -0.0019 \\ 0.0368 \end{bmatrix}$$

$$\boldsymbol{w}_2 = \begin{bmatrix} 0.799 \\ -0.280 \end{bmatrix} + \begin{bmatrix} -0.0019 \\ 0.0368 \end{bmatrix} = \begin{bmatrix} 0.797 \\ -0.244 \end{bmatrix}$$

In Fig. 6.1, the path of gradient descent method with 100 data values of the PDO starting in January 1900 is presented with different learning rates (0.1, 0.01, and 0.001) and the number of epochs (10^4 and 10^5) with the initial point as $w_1 = 4$ and $\alpha_1 = -4$. As shown in the top left panel and the bottom right panel, the learning rate of 0.1 is too large to have appropriate learning. Due to the large learning rate, the learning diverges. With the smaller learning rate as 0.001, the learning can reach the optimum value as shown in the bottom left panel. The gradient descent learning presents that variable α_1 is much more highly updated, followed by w_1 when α_1 is reached at the valley (see the top right panel).

6.2 Adagrad

The adaptive gradient algorithm, termed Adagrad (Duchi et al. 2011), dynamically incorporates the knowledge of the geometry of the data learned from earlier iterations. The algorithm simply decreases the change rate for the parameters that have already been changed much and vice versa. It can be expressed as:

$$\boldsymbol{w}_{t+1} = \boldsymbol{w}_t - \eta \frac{1}{\sqrt{\boldsymbol{h}_t}} \frac{\partial E}{\partial \boldsymbol{w}} \tag{6.4}$$

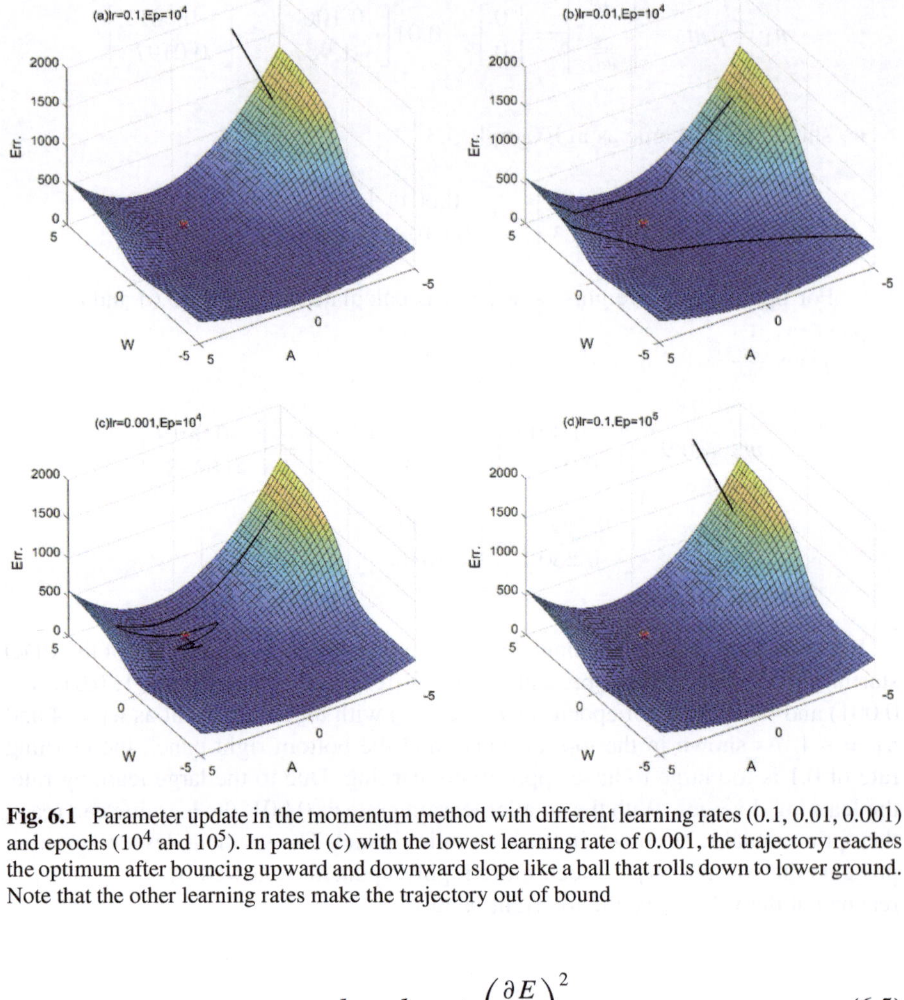

Fig. 6.1 Parameter update in the momentum method with different learning rates (0.1, 0.01, 0.001) and epochs (10^4 and 10^5). In panel (c) with the lowest learning rate of 0.001, the trajectory reaches the optimum after bouncing upward and downward slope like a ball that rolls down to lower ground. Note that the other learning rates make the trajectory out of bound

$$\boldsymbol{h}_t = \boldsymbol{h}_{t-1} + \left(\frac{\partial E}{\partial \boldsymbol{w}}\right)^2 \tag{6.5}$$

where h will increase to infinity if t increases and eventually, no adaptation can be made. In other words, the learning rate, considered as $\eta\frac{1}{\sqrt{\mathbf{h}_t}}$ diminishes and no learning can be made.

The learning rate as $\frac{\eta}{\sqrt{\mathbf{h}_t}}$ is varied at each step and each weight. The weight with a high gradient takes on a lower learning rate from Eq. (6.5).

Example 6.2 Update the weights of the network (i.e., w_1 and α_1) in Fig. 5.1 with the Adagrad method. Here, the current w_1 and α_1 are 0.8 and −0.3, respectively, and the learning rate is 0.01 and employ the dataset as in Table 6.1. Let $\boldsymbol{h}_0 = \begin{bmatrix} 0 \\ 0 \end{bmatrix}$.

Solution:

(1) $y_{p,i}$ is calculated with Eq. (5.6) and $e_i = y_{o,i} - y_{p,i}$.
(2) Then, estimate $\frac{\partial E}{\partial w_1} = 0.106$ and $\frac{\partial E}{\partial \alpha_1} = -1.967$.
(3) For $\boldsymbol{h}_1$ and $\boldsymbol{w}_1$,

$$\boldsymbol{h}_1 = \begin{bmatrix} 0 \\ 0 \end{bmatrix} + \begin{bmatrix} 0.106 \\ -1.967 \end{bmatrix}^2 = \begin{bmatrix} 0.011 \\ 3.869 \end{bmatrix}$$

$$\frac{1}{\sqrt{\boldsymbol{h}_1}} = \begin{bmatrix} 0.508 \\ 9.436 \end{bmatrix}$$

$$\boldsymbol{w}_1 = \begin{bmatrix} 0.8 \\ -0.3 \end{bmatrix} - 0.01 \begin{bmatrix} 0.508 \\ 9.436 \end{bmatrix} \begin{bmatrix} 0.106 \\ -1.967 \end{bmatrix} = \begin{bmatrix} 0.790 \\ -0.290 \end{bmatrix}$$

(4) With re-estimating $y_{p,i}$ and e_i, $\boldsymbol{w}_2$ is estimated as follows:

$\frac{\partial E}{\partial w_1} = 0.102$ and $\frac{\partial E}{\partial \alpha_1} = -1.935$

$$\boldsymbol{h}_2 = \begin{bmatrix} 0.0112 \\ 3.869 \end{bmatrix} + \begin{bmatrix} 0.102 \\ -1.935 \end{bmatrix}^2 = \begin{bmatrix} 0.0216 \\ 7.612 \end{bmatrix}$$

$$\frac{1}{\sqrt{\boldsymbol{h}_1}} = \begin{bmatrix} 6.8084 \\ 0.3625 \end{bmatrix}$$

$$\boldsymbol{w}_2 = \begin{bmatrix} 0.790 \\ -0.290 \end{bmatrix} - 0.01 \begin{bmatrix} 6.8084 \\ 0.3625 \end{bmatrix} \begin{bmatrix} 0.102 \\ -1.935 \end{bmatrix} = \begin{bmatrix} 0.783 \\ -0.283 \end{bmatrix}$$

(5) Repeat step (4).

In Fig. 6.2, the path of the Adagrad method with 100 data values of the PDO starting in January 1900 is presented as the same as in Fig. 6.1. As mentioned above, the learning rate is adjusted according to the accumulated gradient in Eq. (6.5). In contrast to the momentum method increasing the speed, Adagrad reduces the speed for large gradients. This aspect of the Adagrad is presented well in Fig. 6.2. The learning speed is so slow that only the bottom right panel with $\eta = 0.1$ and ephoch $= 10^5$ reaches the optimum, while the others do not reach. However, it seems efficient with no bouncing around, unlike the momentum method.

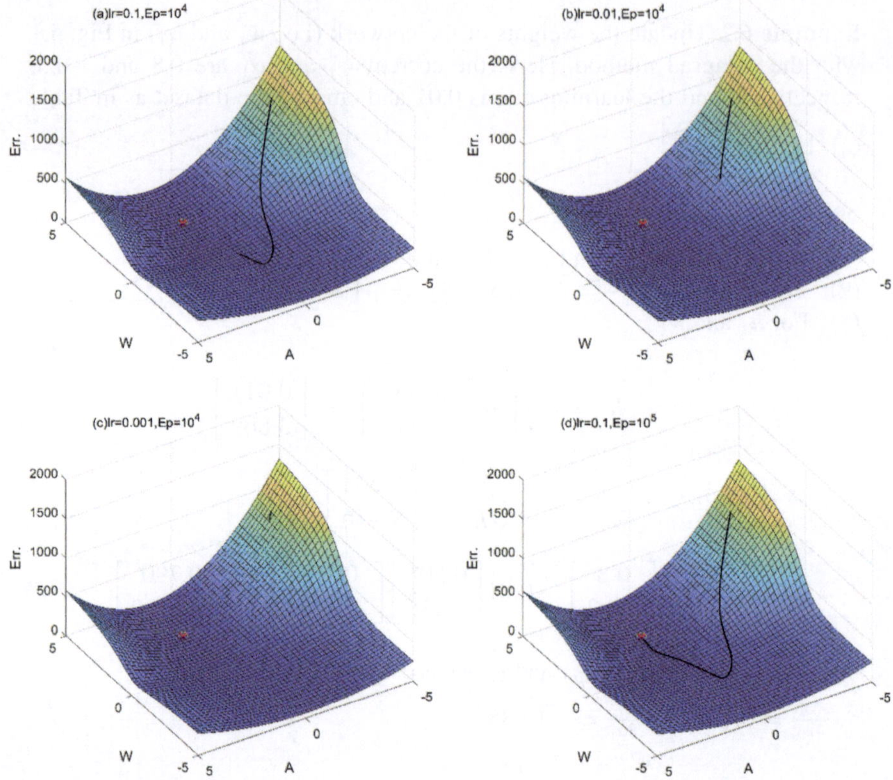

Fig. 6.2 Parameter update for Adagrad with the different learning rates (0.1, 0.01, and 0.001) and epochs (10^4 and 10^5)

6.3 RMSprop

To solve the learning rate diminishing problem in Adagrad, RMSprop (Root Mean Square propagation) has been proposed by Hinton (2012) in his lecture course. The gradient in Adagrad is easy to vanish as its accumulating effect. RMSprop smartly handles this problem by decreasing the step for a large gradient and increasing the step for a small gradient by balancing the step size. It uses a moving average of the squared gradient to normalize the gradient. The method can be described as follows:

$$\boldsymbol{w}_{t+1} = \boldsymbol{w}_t - \eta \frac{1}{\sqrt{\boldsymbol{h}_t + \epsilon}} \frac{\partial E}{\partial \boldsymbol{w}} \tag{6.6}$$

$$\boldsymbol{h}_t = \nu \boldsymbol{h}_{t-1} + (1-\nu)\left(\frac{\partial E}{\partial \boldsymbol{w}}\right)^2 \tag{6.7}$$

where ϵ is prohibited not to become zero (10^{-8}) and ν decreases the effect of the previous influence (e.g., 0.999 or 0.9). By multiplying ν, the effect of a gradient in long time lag to the learning rate diminishes, since ν keeps multiplying as ν^t.

Example 6.3 Update the weights of the network (i.e., w_1 and α_1) in Fig. 5.1 with the RMSprop method. Here, the current w_1 and α_1 are 0.8 and −0.3, respectively, and the learning rate is 0.01 and employ the dataset as in Table 6.1. Let $\boldsymbol{h}_0 = \begin{bmatrix} 0 \\ 0 \end{bmatrix}$ and ϵ and ν be 10^{-8} and 0.9, respectively.

Solution:

(1) $y_{p,i}$ is calculated with Eq. (5.6) and $e_i = y_{o,i} - y_{p,i}$
(2) Then, estimate $\frac{\partial E}{\partial w_1} = 0.106$ and $\frac{\partial E}{\partial \alpha_1} = -1.967$
(3) For $\boldsymbol{h}_1$ and $\boldsymbol{w}_1$

$$\boldsymbol{h}_1 = 0.9\begin{bmatrix} 0 \\ 0 \end{bmatrix} + (1-0.9)\begin{bmatrix} 0.106 \\ -1.967 \end{bmatrix}^2 = \begin{bmatrix} 0.0011 \\ 0.3869 \end{bmatrix}$$

$$\boldsymbol{w}_1 = \begin{bmatrix} 0.8 \\ -0.3 \end{bmatrix} - 0.01\begin{bmatrix} 29.84 \\ 1.61 \end{bmatrix}\begin{bmatrix} 0.106 \\ -1.967 \end{bmatrix} = \begin{bmatrix} 0.768 \\ -0.268 \end{bmatrix}$$

Note that 1/sqrt(0.0011 + 10^{-8}) = 29.84 and 1/sqrt(0.3869 + 10^{-8}) = 1.61.

(4) With reestimating $y_{p,i}$ and e_i, $\boldsymbol{w}_2$ is estimated as follows:
$\frac{\partial E}{\partial w_1} = 0.0926$ and $\frac{\partial E}{\partial \alpha_1} = -1.8650$

$$\boldsymbol{h}_2 = 0.9\begin{bmatrix} 0.0011 \\ 0.3869 \end{bmatrix} + 0.1\begin{bmatrix} 0.0926 \\ -1.8650 \end{bmatrix}^2 = \begin{bmatrix} 0.0019 \\ 0.6960 \end{bmatrix}$$

$$\boldsymbol{w}_2 = \begin{bmatrix} 0.768 \\ -0.268 \end{bmatrix} - 0.01\begin{bmatrix} 23.1304 \\ 1.1986 \end{bmatrix}\begin{bmatrix} 0.0926 \\ -1.8650 \end{bmatrix} = \begin{bmatrix} 0.7469 \\ -0.2460 \end{bmatrix}$$

(5) Repeat step (4).

The detailed results are shown in Fig. 6.3. The RMSprop method has been developed to avoid the learning rate diminishing problem. At the small learning rate of η = 0.1, the trajectory is bouncing around, especially at the final step to the optimum,

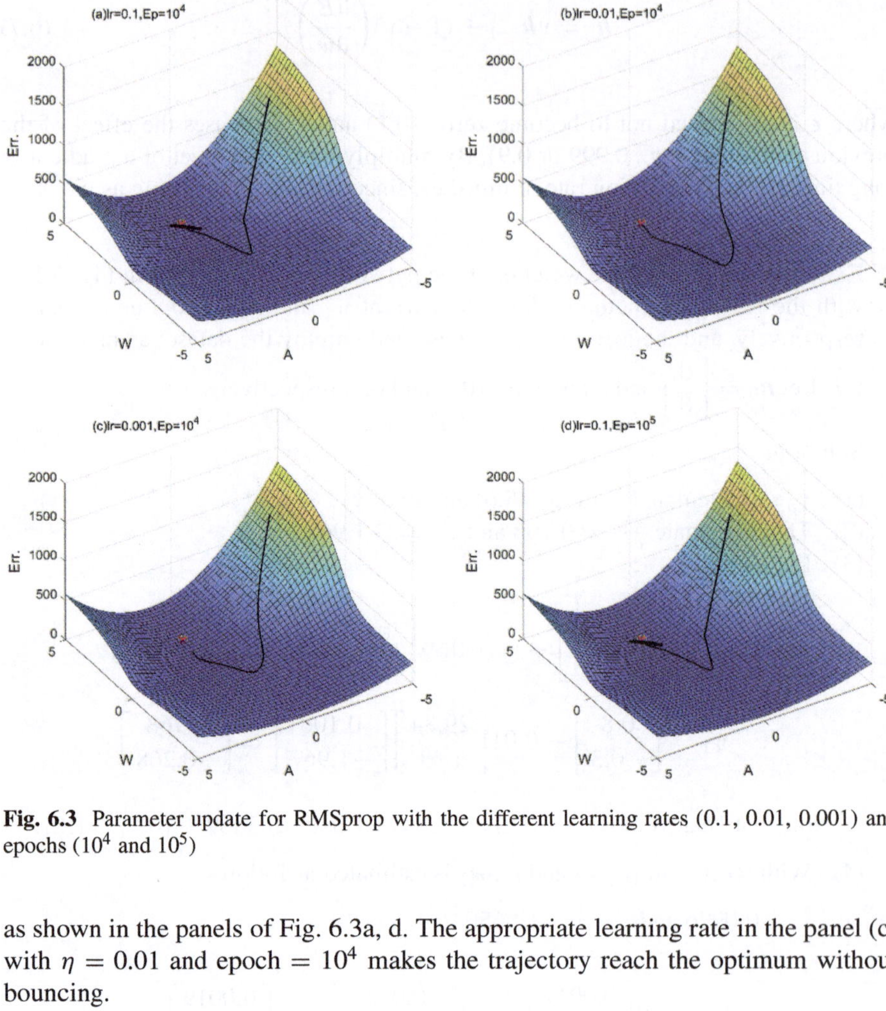

Fig. 6.3 Parameter update for RMSprop with the different learning rates (0.1, 0.01, 0.001) and epochs (10^4 and 10^5)

as shown in the panels of Fig. 6.3a, d. The appropriate learning rate in the panel (c) with $\eta = 0.01$ and epoch $= 10^4$ makes the trajectory reach the optimum without bouncing.

6.4 Adam

Kingma and Ba (2015) proposed an algorithm for first-order gradient-based optimization of stochastic objective functions employing adaptive estimates of the lower-order moment, which is easy to implement and computationally efficient. Adam stands for Adaptive moment estimation. The method calculates each learning rate of weights from estimates of first and second moments of the gradients. It combines two algorithms of Adagrad and RMSprop, but more likely, it is the combination of RMSprop and momentum. Adam is like a ball sliding down on a hill with momentum but with friction force. The method tends to find a minimum solution well, especially in a flat region.

The Adam can be expressed as

$$\boldsymbol{w}_{t+1} = \boldsymbol{w}_t - \eta \frac{1}{\sqrt{\frac{\mathbf{h}_t}{1-\nu^t}} + \epsilon}\left(\frac{\boldsymbol{m}_t}{1-\mu^t}\right) \tag{6.8}$$

$$\boldsymbol{h}_t = \nu \boldsymbol{h}_{t-1} + (1-\nu)\left(\frac{\partial E}{\partial \boldsymbol{w}}\right)^2 \tag{6.9}$$

$$\boldsymbol{m}_t = \mu \boldsymbol{m}_{t-1} + (1-\mu)\frac{\partial E}{\partial \boldsymbol{w}} \tag{6.10}$$

where μ and ν are the hyperparameters generally set as 0.999.

Adam uses initialization bias terms by deriving the term for the second moment estimate as follows. Equation (6.10) can be simplified as

$$\boldsymbol{m}_t = \mu \boldsymbol{m}_{t-1} + (1-\mu)\boldsymbol{g}_t \tag{6.11}$$

where $\boldsymbol{g}_t = \left(\frac{\partial E}{\partial \boldsymbol{w}}\right)_t$. It is explicitly expressed as:

$$\boldsymbol{m}_1 = (1-\mu)\boldsymbol{g}_1 \tag{6.12}$$

$$\boldsymbol{m}_2 = \mu(1-\mu)\boldsymbol{g}_1 + (1-\mu)\boldsymbol{g}_2 \tag{6.13}$$

$$\boldsymbol{m}_3 = \mu \boldsymbol{m}_2 + (1-\mu)\boldsymbol{g}_3 = \mu^2(1-\mu)\boldsymbol{g}_1 + \mu(1-\mu)\boldsymbol{g}_2 + (1-\mu)\boldsymbol{g}_3 \tag{6.14}$$

It can be rewritten as

$$\boldsymbol{m}_t = (1-\mu)\sum_{k=1}^{t} \mu^{t-k}\boldsymbol{g}_k \tag{6.15}$$

Here, $\sum_{k=1}^{t} \mu^{t-k} = \sum_{k=0}^{t-1} \mu^k = \frac{1-\mu^t}{1-\mu}$.
Then,

$$\mathrm{E}[\boldsymbol{m}_t] = E[\boldsymbol{g}_t](1-\mu)\sum_{k=1}^{t} \mu^{t-k} + \psi = E[\boldsymbol{g}_t](1-\mu^t) + \psi \tag{6.16}$$

Finally, the bias of $\boldsymbol{m}_t$ is $\left(1-\mu^t\right)$. Therefore, the updating rule in Eq. (6.8) applies the bias for unbiased estimation. The same implication is made for $\boldsymbol{h}_t$.

Example 6.4 Update the weights of the network (i.e., w_1 and α_1) in Fig. 5.1 with the Adam method. Here, the current w_1 and α_1 are 0.8 and −0.3, respectively, and the learning rate is 0.01 and employ the dataset as in Table 6.1. Let $\boldsymbol{h}_0 = \boldsymbol{m}_0 = \begin{bmatrix} 0 \\ 0 \end{bmatrix}$ and ϵ and ν be $10^{-8}8$ and 0.9, respectively, and μ be 0.9.

Solution:

(1) $y_{p,i}$ is calculated with Eq. (5.6) and $e_i = y_{o,i} - y_{p,i}$.
(2) Then, estimate $\frac{\partial E}{\partial w_1} = 0.106$ and $\frac{\partial E}{\partial \alpha_1} = -1.967$.
(3) For $\boldsymbol{h}_1$ and $\boldsymbol{w}_1$

$$\boldsymbol{h}_1 = 0.9\begin{bmatrix} 0 \\ 0 \end{bmatrix} + (1-0.9)\begin{bmatrix} 0.106 \\ -1.967 \end{bmatrix}^2 = \begin{bmatrix} 0.0011 \\ 0.3869 \end{bmatrix}$$

$$\boldsymbol{m}_1 = 0.9\begin{bmatrix} 0 \\ 0 \end{bmatrix} + (1-0.9)\begin{bmatrix} 0.106 \\ -1.967 \end{bmatrix} = \begin{bmatrix} 0.0106 \\ -0.1967 \end{bmatrix}$$

Note that for $i = 0$, use $1-\mu^i = 1$ and $1-\nu^i = 1$.

$$\boldsymbol{w}_1 = \boldsymbol{w}_0 - \eta\frac{1}{\sqrt{\mathbf{h}_1} + \epsilon}(\boldsymbol{m}_1) = \begin{bmatrix} 0.8 \\ -0.3 \end{bmatrix} - 0.01\begin{bmatrix} 29.84 \\ 1.61 \end{bmatrix}\begin{bmatrix} 0.0106 \\ -0.1967 \end{bmatrix} = \begin{bmatrix} 0.768 \\ -0.268 \end{bmatrix}$$

(4) With re-estimated $y_{p,i}$ and e_i, $\boldsymbol{w}_2$ is estimated as follows:
$\frac{\partial E}{\partial w_1} = 0.0926$ and $\frac{\partial E}{\partial \alpha_1} = -1.8650$

$$\boldsymbol{h}_2 = 0.9\begin{bmatrix} 0.0011 \\ 0.3869 \end{bmatrix} + 0.1\begin{bmatrix} 0.0926 \\ -1.8650 \end{bmatrix}^2 = \begin{bmatrix} 0.0019 \\ 0.6960 \end{bmatrix}$$

$$\boldsymbol{m}_2 = 0.9\begin{bmatrix} 0.0106 \\ -0.1967 \end{bmatrix} + (1-0.9)\begin{bmatrix} 0.0926 \\ -1.8650 \end{bmatrix} = \begin{bmatrix} 0.0188 \\ -0.3635 \end{bmatrix}$$

$$w_2 = w_1 - \eta\frac{1}{\sqrt{\frac{\mathbf{h}_1}{1-\nu^1} + epsilon}}\left(\frac{m_1}{1-\mu^1}\right) = \begin{bmatrix} 0.768 \\ -0.268 \end{bmatrix} - 0.01\begin{bmatrix} 10.0823 \\ 0.5225 \end{bmatrix}\begin{bmatrix} 0.0990 \\ -1.9133 \end{bmatrix} = \begin{bmatrix} 0.7584 \\ -0.2454 \end{bmatrix}$$

(5) Repeat step (4).

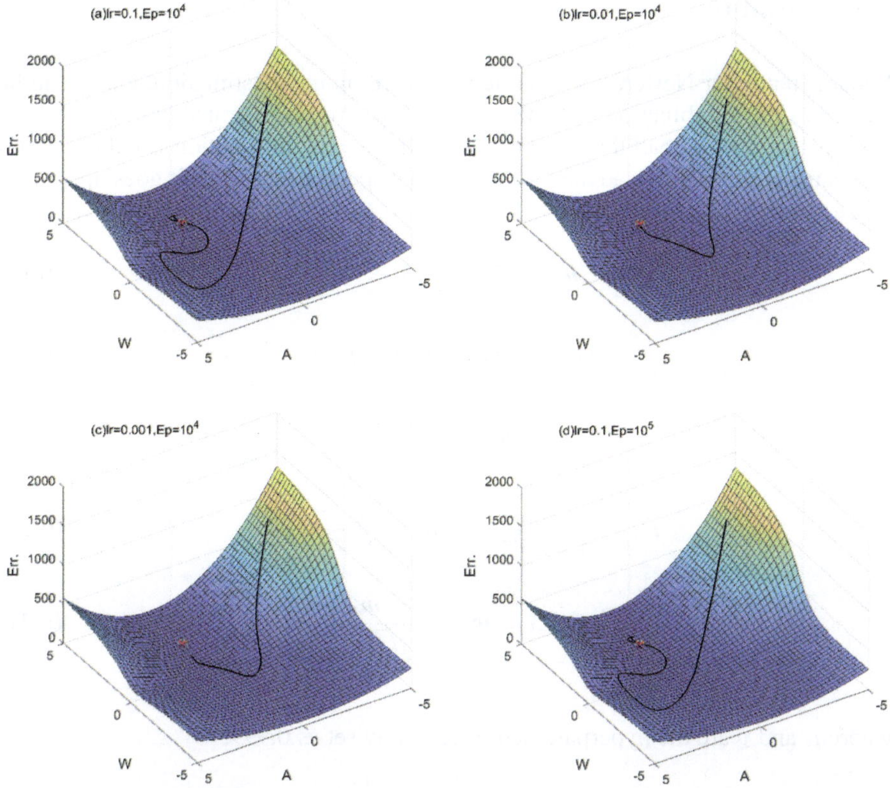

Fig. 6.4 Parameter update with Adam with different learning rates (0.1, 0.01, 0.001) and epochs (10^4 and 10^5)

The Adam method is applied to the same dataset as in Fig. 6.1, and the result is shown in Fig. 6.4. At a relatively large learning rate of $\eta = 0.1$ in panel (a) and (d), the trajectory rolls around but to a lower degree compared to the momentum method similar to the RMSprop in Fig. 6.3. All the learning rates were tested to reach the optimum except the one with $\eta = 0.001$ in panel (c). As noticed, its speed is relatively adjusted well to find the optimum by combining Adagrad and RMSprop. Adagrad finds an efficient route to reach the optimum, while RMSprop adjusts its momentum speed not to lose an efficient route.

6.5 Nadam

Nadam stands for Nesterov accelerated adaptive moment estimation, and it can be viewed as the combination of RMSprop and Nesterov accelerated gradient (NAG). NAG is like a smart ball that goes down a hill but gets information whether the hill slopes up again or not in advance and slows its speed down when it goes up.

The Nadam is expressed as

$$\boldsymbol{h}_t = \nu \boldsymbol{h}_{t-1} + (1-\nu)\boldsymbol{g}_t^2 \tag{6.17}$$

$$\boldsymbol{m}_t = \mu_t \boldsymbol{m}_{t-1} + (1-\mu_t)\boldsymbol{g}_t \tag{6.18}$$

$$\hat{\boldsymbol{h}}_t = \frac{\nu \boldsymbol{h}_t}{1-\nu^t} \tag{6.19}$$

$$\hat{\boldsymbol{m}}_t = \frac{\mu_{t+1}}{1-\prod_{j=1}^{t+1}\mu_j}\boldsymbol{m}_t + \frac{(1-\mu_t)\boldsymbol{g}_t}{1-\prod_{j=1}^{t}\mu_j} \tag{6.20}$$

$$\boldsymbol{w}_{t+1} = \boldsymbol{w}_t - \eta \frac{\hat{\boldsymbol{m}}_t}{\sqrt{\hat{\boldsymbol{h}}_t} + \epsilon} \tag{6.21}$$

where μ and ν are the hyperparameters generally set as 0.999.

Example 6.5 Update the weights of the network (i.e., w_1 and α_1) in Fig. 5.1 with the Nadam method. Here, the current w_1 and α_1 are 0.8 and −0.3, respectively, and the learning rate is 0.01 and employ the dataset as in Table 6.1. Let $\boldsymbol{h}_0 = \boldsymbol{m}_0 = \begin{bmatrix} 0 \\ 0 \end{bmatrix}$ and ϵ and ν be 10^{-8} and 0.9, respectively and μ be 0.9.

Solution:

Steps (1)–(2) are the same as Adam. Therefore,

(3) $\boldsymbol{w}_1$ is estimated as follows:

$$\boldsymbol{h}_1 = \begin{bmatrix} 0.0011 \\ 0.3869 \end{bmatrix}, \boldsymbol{m}_1 = \begin{bmatrix} 0.0106 \\ -0.1967 \end{bmatrix}$$

$$\hat{\boldsymbol{h}}_1 = \frac{\nu \boldsymbol{h}_1}{1-\nu^1} = \begin{bmatrix} 0.0101 \\ 3.4821 \end{bmatrix}$$

$$\hat{\boldsymbol{m}}_1 = \frac{\mu_{t+1}}{1-\prod_{j=1}^{t+1}\mu_j}\boldsymbol{m}_t + \frac{(1-\mu_t)\boldsymbol{g}_t}{1-\prod_{j=1}^{t}\mu_j} = \frac{\mu}{1-\mu^2}\begin{bmatrix} 0.0106 \\ -0.1967 \end{bmatrix} + \frac{1-\mu}{1-\mu^1}\begin{bmatrix} 0.106 \\ -1.967 \end{bmatrix} = \begin{bmatrix} 0.1562 \\ -2.8987 \end{bmatrix}$$

$$w_1 = w_0 - \eta \frac{\hat{m}_1}{\sqrt{\hat{h}_1 + \epsilon}} = \begin{bmatrix} 0.8 \\ -0.3 \end{bmatrix} - 0.01 \begin{bmatrix} 1.5534 \\ -1.5534 \end{bmatrix} = \begin{bmatrix} 0.7845 \\ -0.2845 \end{bmatrix}$$

(4) With reestimated $y_{p,i}$ and e_i, w_2 is estimated as follows:

$\frac{\partial E}{\partial w_1} = 0.0994$ and $\frac{\partial E}{\partial \alpha_1} = -1.9168$.

$h_2 = \begin{bmatrix} 0.0020 \\ 0.7156 \end{bmatrix}$ and $m_2 = \begin{bmatrix} 0.0197 \\ -0.3687 \end{bmatrix}$

$$\hat{h}_2 = \frac{\nu h_2}{1 - \nu^2} = \begin{bmatrix} 0.0067 \\ 2.3892 \end{bmatrix}$$

$$\hat{m}_1 = \frac{\mu}{1 - \mu^3} \begin{bmatrix} 0.0106 \\ -0.1967 \end{bmatrix} + \frac{1 - \mu}{1 - \mu^2} \begin{bmatrix} 0.0994 \\ -1.9168 \end{bmatrix} = \begin{bmatrix} 0.1170 \\ -2.2333 \end{bmatrix}$$

$$w_2 = w_1 - \eta \frac{\hat{m}_1}{\sqrt{\hat{h}_1 + \epsilon}} = \begin{bmatrix} 0.7845 \\ -0.2845 \end{bmatrix} - 0.01 \begin{bmatrix} 1.2023 \\ -1.2130 \end{bmatrix} = \begin{bmatrix} 0.7724 \\ -0.2723 \end{bmatrix}$$

(5) Repeat step (4).

$$w_2 = w_1 - \eta \frac{1}{\sqrt{\frac{\mathbf{h}_1}{1 - \nu^1} + \epsilon}} \left(\frac{\mu m_{t-1}}{1 - \mu^i} + \frac{(1 - \mu) g_t}{1 - \mu^1} \right)$$

$$= \begin{bmatrix} 0.768 \\ -0.268 \end{bmatrix} - 0.01 \begin{bmatrix} 10.0823 \\ 0.5225 \end{bmatrix} \begin{bmatrix} 0.0990 \\ -1.9133 \end{bmatrix} = \begin{bmatrix} 0.7584 \\ -0.2454 \end{bmatrix}$$

(6) Repeat step (4).

As shown in the example of the Adam method in Fig. 6.4, the Nadam is applied to the data of 100 scaled monthly PDO values as in Example 6.5, and the result is presented in Fig. 6.5. The difference between Adam and Nadam is clearly presented as the Nadam finds the optimum in a very efficient manner as in panel (a), (b), and (d). However, Nadam is slightly slower than Adam, as presented in panel (c) of Fig. 6.5 compared to panel (c) of Fig. 6.4.

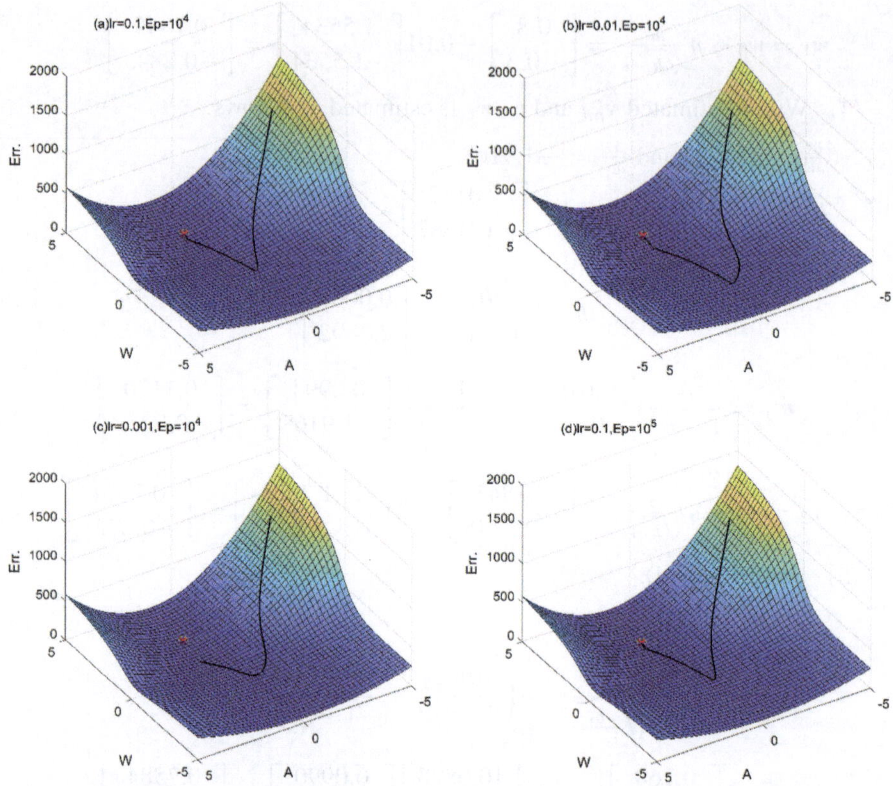

Fig. 6.5 Parameter update for Nadam with the different learning rates (0.1, 0.01, 0.001) and epochs (10^4 and 10^5)

6.6 Python Coding of Updating Weights

The following presents the python code in all the examples in this chapter

```
[Python 6.1] Program the learning algorithms in Chapter 6
import numpy as np
import pandas as pd
import matplotlib.pyplot as plt

#####################################################
# Example 1 parameter estimation for Simple Neural Network
# with Sigmoid activation function
def update_para(Ypred,Yobs,X,W,A,lr):
    W -= lr*dE_dW(Ypred,Yobs,X,W,A)
    A -= lr*dE_dA(Ypred,Yobs,X,A)
    return W, A
```

```
def update_para_Momentum(Ypred,Yobs,X,W,A,lr,mw,ma):
    mu=0.9
    dEW=dE_dW(Ypred,Yobs,X,W,A)
    mw=mu*mw-lr*dEW
    W +=mw
    dEA=dE_dA(Ypred,Yobs,X,A)
    ma=mu*ma-lr*dEA
    A +=ma
    return W, A, mw, ma

def update_para_AdaGrad(Ypred,Yobs,X,W,A,lr,hw,ha):
    v=0.999
    dEW=dE_dW(Ypred,Yobs,X,W,A)
    hw += dEW**2
    W -= lr*dEW/(hw)**0.5
    dEA=dE_dA(Ypred,Yobs,X,A)
    ha += dEW**2
    A -= lr*dEA/(ha)**0.5
    return W, A, hw, ha
```

```
def update_para_RMSprop(Ypred,Yobs,X,W,A,lr,hw,ha):
    v=0.999
    dEW=dE_dW(Ypred,Yobs,X,W,A)
    hw = v*hw+(1-v)*dEW**2
    W -= lr*dEW/(hw+10**(-8))**0.5
    dEA=dE_dA(Ypred,Yobs,X,A)
    ha = v*ha+(1-v)*dEA**2
    A -= lr*dEA/(ha+10**(-8))**0.5
    return W, A, hw, ha
```

```
[Python 6.1] Continue
def update_para_Adam(Ypred,Yobs,X,W,A,lr,hw,ha,mw,ma,count):
    mu=0.9
    v=0.999
    dEW=dE_dW(Ypred,Yobs,X,W,A)
    hw = v*hw+(1-v)*dEW**2
    mw = mu*mw+(1-mu)*dEW
    W -= lr*(mw/(1-mu**count))/(hw/(1-v**count)+10**(-8))**0.5

    dEA=dE_dA(Ypred,Yobs,X,A)
    ha = v*ha+(1-v)*dEA**2
    ma = mu*ma+(1-mu)*dEA
    A -= lr*(ma/(1-mu**count))/(ha/(1-v**count)+10**(-8))**0.5
    return W, A, hw, ha, mw, ma
```

```
def update_para_Nadam(Ypred,Yobs,X,W,A,lr,hw,ha,mw,ma,count):
    mu=0.9
    v=0.999
    eps=10**(-8)
    dEW=dE_dW(Ypred,Yobs,X,W,A)
    dEA=dE_dA(Ypred,Yobs,X,A)
    dE=[dEW,dEA]
    WA=[W,A]
    h = v*h+(1-v)*dE**2
    m = mu*m+(1-mu)*dE
    hh = v*h/(1-v**count)
```

```
mh = mu/(1-mu**(count+1))*m+(1-mu)*dE/(1-mu**t)
WA -= lr*mh/((hh)**0.5+eps)
W =WA[0]
A =WA[1]
hw=h[0]
ha=h[1]
mw=m[0]
ma=m[1]

return W, A, hw, ha, mw, ma
```

6.7 Exercises

1. Perform the parameter updates ($t = 2$) for the network in Fig. 5.3 to the data in Table 5.2 using the following algorithms:

   (a) Momentum ($\mu = 0.9$, $\eta = [0.1, 0.01, \text{and } 0.001]\mu\nu$)
   (b) Adagrad ($\mu = 0.9$, $\eta = [0.1, 0.01, \text{and } 0.001]\mu\nu$)
   (c) RMSprop ($\mu = 0.9$, $\eta = [0.1, 0.01, \text{and } 0.001]$, $\varepsilon = 10^{-8}$, $\nu = 0.9$)
   (d) Adam ($\mu = 0.9$, $\eta = [0.1, 0.01, \text{and } 0.001]$, $\varepsilon = 10^{-8}$, $\nu = 0.9$)
   (e) Nadam ($\mu = 0.9$, $\eta = [0.1, 0.01, \text{and } 0.001]$, $\varepsilon = 10^{-8}$, $\nu = 0.9$)

2. Program the above with Python by employing the function in Python 6.1.

References

Duchi J, Hazan E, Singer Y (2011) Adaptive subgradient methods for online learning and stochastic optimization. J Mach Learn Res 12:2121–2159

Hinton GE (2012) Neural networks for machine learning. Coursera

Kingma DP, Ba JL (2015) Adam: a method for stochastic optimization. In: 3rd international conference for learning representations, San Diego, pp 1–15

Qian N (1999) On the momentum term in gradient descent learning algorithms. Neural Netw 12(1):145–151. https://doi.org/10.1016/S0893-6080(98)00116-6

Chapter 7
Improving Model Performance

Abstract In order improve the performance of a neural network model, a number of ways have been studied. In this chapter, minibatch and k-fold cross-validation are explained. The basic idea of these two methods is on controlling the dataset, since repeated usage of the same dataset for training and validation might result in overfitting. Furthermore, regularization of the neural network model training by L-norm regularization and dropout of hidden nodes are explained in this chapter to avoid overfitting.

7.1 Batching and Minibatch

For updating weights in Eq. (5.2), the gradient descent algorithm is essential and its derivative is estimated with the whole training dataset. This is called "batch gradient descent," that is, employing all the training dataset. However, in case the training dataset is enormous, it is very time-consuming, even just for a single step. To resolve this problem, stochastic gradient descent (SGD) was suggested in a way that only one training value is used at a time. In this case, a target loss function fluctuates over the training dataset, but it will not necessarily decrease the time, even though the loss function eventually decreases.

In order to minimize this fluctuation and long run to learn the whole training dataset, minibatch gradient descent was proposed as a mixture of batch gradient descent and SGD. In minibatch gradient descent, a batch with a fixed number of training datasets is used to update the weights of the network. Note that one iteration that takes to train the network with the whole training dataset through minibatch gradient descent is called an **epoch**.

T. Lee et al., *Deep Learning for Hydrometeorology and Environmental Science*,
Water Science and Technology Library 99,
https://doi.org/10.1007/978-3-030-64777-3_7

7.2 Validation

Through training, the target loss function decreases always. However, it results in overfitting that is too much fitting to the employed data, called overfitting. Therefore, a portion of the dataset is separated for an unbiased evaluation of the model in tuning hyperparameters.

Furthermore, cross-validation is used as a resampling procedure (i.e., randomly selecting a portion of the dataset with replacement) to evaluate its performance on limited data. With a specific value as *k*, the dataset is split into training and validation datasets randomly. The procedure of k-fold cross-validation (KfCV) is

(1) Shuffle the whole dataset with *N* record length.
(2) Split the dataset into k groups and each group has *N*/*k* record length.
(3) Hold out one group among *k* groups and fit the network with the remaining *k*-1 groups.
(4) Evaluate the fitted network with the hold group.
(5) Repeat steps of (3) and (4) for *k* times.

7.2.1 Python Coding of K-Fold Cross-Validation

The K-fold cross-validation is performed with two functions of Kfold_dataXY_DeepLearning and CrossVald_Kfold. The first function is to prepare the dataset and the second function is to validate its dataset iteratively for *k* times.

```
[Python 7.1] Validation code
# Function to create the Kfold datasets
def Kfold_dataXY_DeepLearning(DataX,DataY,Nfolds):
    Ndata=len(DataX)
    N=(Ndata//Nfolds)
    ns=np.linspace(0,(Ndata//Nfolds)*Nfolds,Nfolds+1).astype(int)

    Xtrain_kfd=[[] for i in range(len(ns)-1)]
    Ytrain_kfd=[[] for i in range(len(ns)-1)]
    Xtest_kfd=[[] for i in range(len(ns)-1)]
    Ytest_kfd=[[] for i in range(len(ns)-1)]

    for isd in range(len(ns)-1):
        testX_org=DataX[ns[isd]:ns[isd+1]-1,:]
        Xtest_kfd[isd] = np.reshape(testX_org, (testX_org.shape[0], 1, testX_org.shape[1]))
        Ytest_kfd[isd]=DataY[ns[isd]:ns[isd+1]-1]

        trainX_org=np.delete(DataX,range(ns[isd],ns[isd+1]-1),0)
        Xtrain_kfd[isd]= np.reshape(trainX_org, (trainX_org.shape[0], 1, trainX_org.shape[1]))
        Ytrain_kfd[isd]=np.delete(DataY,range(ns[isd],ns[isd+1]-1))

    return Xtrain_kfd, Ytrain_kfd, Xtest_kfd, Ytest_kfd
```

7.3 Regularization

Regularization is a set of techniques to prevent overfitting by adding information in neural networks. Among others, L1, L2, and dropout techniques are discussed.

7.3.1 L-Norm Regularization

Norm is to measure a distance between vectors and can be denoted as

$$\|\mathbf{x}\|_p = \left(\sum_{i=1}^{N} |x_i|^p\right)^{1/p} \tag{7.1}$$

where $p \geq 1$ and a real number. If $p = 1$ (i.e., $L1$ norm) for a vector distance

$$d(\mathbf{x}, \mathbf{y})_1 = \|\mathbf{x} - \mathbf{y}\|_1 = \sum_{i=1}^{N} |x_i - y_i| \tag{7.2}$$

For $p = 1$, it becomes the Euclidean norm as

$$d(\mathbf{x}, \mathbf{y})_2 = \|\mathbf{x} - \mathbf{y}\|_2 = \sqrt{\sum_{i=1}^{N} (x_i - y_i)^2} \tag{7.3}$$

Also, the L2 loss function is described in Eq. (4.12).

Regularization is to lower the complexity of neural networks in training data. The L1 and L2 regularization can be executed by including additional term in the loss function of an ANN model as for L1 regularization,

$$Cost = \frac{1}{N}\sum_{i=1}^{N}\left\{L\left(y_{o,i}, -y_{p,i}\right) + \frac{\lambda}{2}|\mathbf{w}|\right\} \tag{7.4}$$

with and for L2 regularization,

$$Cost = \frac{1}{N}\sum_{i=1}^{N}\left\{L\left(y_{o,i}, -y_{p,i}\right) + \frac{\lambda}{2}|\mathbf{w}|^2\right\} \tag{7.5}$$

where λ is the hyperparameter that controls the importance of the regularization term. Note that the added norm (i.e., $\frac{\lambda}{2}|\mathbf{w}|$ and $\frac{\lambda}{2}|\mathbf{w}|^2$) increases its cost with high values of the weights.

Eventually, the gradient descent in Eq. (5.2) for L1 and L2 regularization becomes

$$\mathbf{w}^{t+1} = \mathbf{w}^{t} - \eta\left[\frac{\partial L(\mathbf{w})}{\partial \mathbf{w}} + \frac{\lambda}{N}\mathrm{sgn}(\mathbf{w_t})\right] \tag{7.6}$$

$$\mathbf{w}^{t+1} = \mathbf{w}^{t} - \eta\left[\frac{\partial L(\mathbf{w})}{\partial \mathbf{w}} + \frac{\lambda}{N}\mathbf{w_t}\right] \tag{7.7}$$

Example 7.1 Update the weights of the network (i.e. w_1 and α_1) in Fig. 5.1 with the R1 and R2 regularization method (set λ as 0.5). Here, the current w_1 and α_1 are 0.8 and −0.3, respectively, and the learning rate is 0.01 and employ the dataset in the 2nd and 3rd columns of Table 6.1.

Solution:

As shown in Table 6.1, $\frac{\partial E}{\partial w_1} = 0.106$ and $\frac{\partial E}{\partial \alpha_1} = -1.967$

For L1 regularization,

$$\boldsymbol{w}^{t+1} = \boldsymbol{w}^{t} - \eta\left[\tfrac{\partial L(\mathbf{w})}{\partial \mathbf{w}} + \tfrac{\lambda}{N}\mathrm{sgn}(\boldsymbol{w}_t)\right]$$

$$\boldsymbol{w}_1 = \begin{bmatrix} 0.8 \\ -0.3 \end{bmatrix} - 0.01\left\{\begin{bmatrix} 0.106 \\ -1.967 \end{bmatrix} + \tfrac{0.5}{10}\begin{bmatrix} +1 \\ -1 \end{bmatrix}\right\} = \begin{bmatrix} 0.798 \\ -0.281 \end{bmatrix}$$

For L2 regularization

$$\boldsymbol{w}^{t+1} = \boldsymbol{w}^{t} - \eta\left[\tfrac{\partial L(\mathbf{w})}{\partial \mathbf{w}} + \tfrac{\lambda}{N}\boldsymbol{w}_t\right]$$

$$\boldsymbol{w}_1 = \begin{bmatrix} 0.8 \\ -0.3 \end{bmatrix} - 0.01\left\{\begin{bmatrix} 0.106 \\ -1.967 \end{bmatrix} + \tfrac{0.5}{10}\begin{bmatrix} 0.8 \\ -0.3 \end{bmatrix}\right\} = \begin{bmatrix} 0.799 \\ -0.280 \end{bmatrix}$$

7.3.2 *Dropout*

Dropout is a simple regularization technique that some of the hidden layer outputs are randomly dropped out during the training of the network. The word "dropout" implies that some portion of the hidden neurons in a network is turned off during training, as shown in Fig. 7.1. Dropout can prevent overfitting and provide ensemble neural networks with different architectures efficiently. The idea of dropout is that a hidden unit trained with dropout must learn to work with a randomly selected unit, and this scheme makes each hidden unit more robust and forces to create useful features on its own without relying on other units to correct its mistakes (Srivastava et al. 2014).

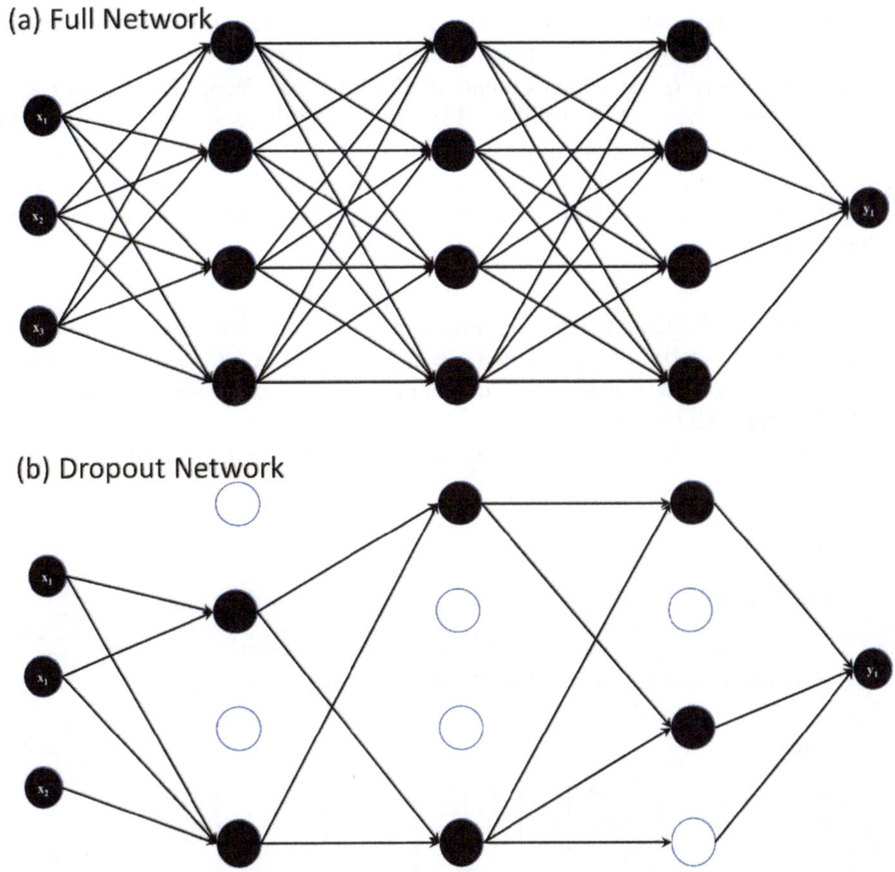

Fig. 7.1 Full and dropout network with three inputs and one output as well as three hidden layers and four hidden nodes in each hidden layer. The emptied circles in panel (**b**) indicate the dropped units

With dropout, the feedforward operation in Eqs. (4.24) and (4.25) with one hidden layer becomes

$$r^j \sim \text{Bernoulli}(p) \text{ for } j = 1, \ldots, nH \tag{7.8}$$

$$\tilde{\boldsymbol{h}}_i = \boldsymbol{h}_i \otimes \mathbf{r} \tag{7.9}$$

$$\boldsymbol{y}_i = g_{HO}\Big(\mathbf{W}_{HO}\tilde{\boldsymbol{h}}_i + b_{HO}\Big) \tag{7.10}$$

where r^j is simulated from Bernoulli distribution, $\mathbf{r}$ is the vector of r^j, $\otimes$ represents the element multiplication, and $\boldsymbol{h}_i$ is already defined in Eq. (4.25). The equation presents that some hidden units are randomly set to zero during forward propagation. The dropout is employed only in training but not in testing.

The probability (p) at which outputs of the layer are dropped out works as a tuning hyperparameter. A probability of 0.5 is commonly used. Note that dropout is not employed after training to predict with the fitted network. Note that Fig. 7.1 employs its probability as 0.5 and half of the nodes are dropped out randomly. Also, one can randomly remove input variables with less probability to dropout between 0.0 and 0.5.

Example 7.2 Execute the dropout for the weights of the network in Example 4.6 as the network consists of two input nodes, one output node and three hidden nodes. Now, employ the same input value and parameters in Example 4.6 Estimate the output (Yi).

Among three hidden nodes, simulate r^j with Bernoulli distribution with probability $p = 0.67$

$r^j \sim \text{Bernoulli}(0.67) \text{for} j = 1, \ldots, 3$

$$\mathbf{r} = \begin{bmatrix} 1 \\ 0 \\ 1 \end{bmatrix}$$

Note that dropout is the random procedure so that dropout hidden nodes are not fixed but randomly selected even with the probability of 0.67.

Then

$$\tilde{\boldsymbol{h}}_i = \boldsymbol{h}_i \otimes \mathbf{r} = \begin{bmatrix} 0.20 \\ 0.62 \\ 0.70 \end{bmatrix} \otimes \begin{bmatrix} 1 \\ 0 \\ 1 \end{bmatrix} = \begin{bmatrix} 0.20 \\ 0 \\ 0.70 \end{bmatrix}$$

$$\left[y_i^1\right] = g_{HO}\left\{\left[-0.363\ -0.252\ 0.859\ 0.656\right]\begin{bmatrix} 1 \\ 0.2 \\ 0 \\ 0.7 \end{bmatrix}\right\} =$$

$g_{HO}\{0.0458\} = 0.0458$

Note that the estimate y_i is smaller than the one from Example 4.6, since the corresponding parameter α_2 is as high as 0.859. This indicates that the dropout algorithm deteriorates one parameter to have a high value by ignoring the hidden state with zero value.

7.3.3 *Python Coding of Regularization*

This is the program for a regularization function for L1 and L2 regularization

```
[Python 7.2] Program the learning algorithms in Chapter 6
import numpy as np
import pandas as pd
import matplotlib.pyplot as plt

##############################################################
# Example 1 parameter estimation for Simple Neural Network
# with Sigmoid activation function
def update_para_L1(Ypred,Yobs,X,W,A,lr):
    W -= lr*(dE_dW(Ypred,Yobs,X,W,A) + lambda*np.sign(W)/len(Yobs))
    A -= lr*(dE_dA(Ypred,Yobs,X,A) +lambda* np.sign(W)/len(Yobs))
    return W, A

def update_para_L2(Ypred,Yobs,X,W,A,lr):
    W -= lr*(dE_dW(Ypred,Yobs,X,W,A)+lambda*W/len(Yobs))
    A -= lr*(dE_dA(Ypred,Yobs,X,A) +lambda*W/len(Yobs))
    return W, A
```

The following is the program for dropout.

```
[Python 7.3] Program the dropout
def dropout(Xd, p_drop=0):
    mask = np.random.uniform(0, 1, Xd.shape) > p_drop
    return mask
```

7.4 Exercises

1. Perform KfCV ($k = 2$) for the network in Fig. 5.3 with the method and the dataset in Figure 5.3 and Table 5.2.
2. Update the weights of the network in Fig. 5.3 with the R1 and R2 regularization method (set λ as 0.5). Apply the same input and output data as in Example 5.4.
3. Execute the dropout for the weights of the network in Example 4.6 as the network consists of two input nodes, one output node and three hidden nodes, but employing tanh function for the hidden activation function. See Example 7.2 for detail.

   (a) Estimate the output (y_i).

   (b) Employ $\mathbf{r} = \begin{bmatrix} 0 \\ 0 \\ 1 \end{bmatrix}$ and $\mathbf{r} = \begin{bmatrix} 0 \\ 1 \\ 1 \end{bmatrix}$ and estimate the hidden state and output.

Reference

Srivastava N, Hinton G, Krizhevsky A, Sutskever I, Salakhutdinov R (2014) Dropout: a simple way to prevent neural networks from overfitting. J Mach Learn Res 15:1929–1958

Chapter 8
Advanced Neural Network Algorithms

Abstract In this chapter, recently developed neural network algorithms are introduced, including extreme learning machine and autoencoder. These algorithms are popularly adopted in deep learning models. Particularly, autoencoder can be used in shrinking the dimension of inputs and outputs as well as hidden nodes.

8.1 Extreme Learning Machine (ELM)

An extreme learning machine (ELM) is an algorithm to train feedforward neural networks. It was first developed for single hidden layer feedforward neural networks (SLFNs) but then extended to multilayer perceptrons as well. For decades, gradient descent-based techniques like backpropagation have been of interest to train neural networks. However, these algorithms generally are slow due to their iterative nature and many learning steps that should be taken in order to achieve acceptable learning performance. ELM was introduced by Huang et al. (2006) to address the time efficiency and the generalization problem of ANN learning methods. Its distinctive characteristic is the ability to train an SLFN analytically in just one single step, which reduces the training time up to 6 orders of magnitude. This remarkable property also makes ELM implementation easier, more efficient, and robust for online and real-time applications and also for processing huge data sets.

In ELM, the randomly initiated input weights and biases remain fixed, and they are not required to be adjusted further. The output weights are determined simply in a linear system of output weights. The idea of "randomness" of hidden nodes makes the hidden weights completely independent of training data and also the parameters of the output layer.

The architecture of an SLFN with d input nodes and L hidden nodes, subject to ELM, is demonstrated in Fig. 8.1. A training set, including a target vector $\mathbf{Y} = [y_1, y_2, \ldots, y_N]^T$ and a feature matrix $\mathbf{X} = [\mathbf{x}_1, \mathbf{x}_2, \ldots, \mathbf{x}_N]^T$ in which $\mathbf{x}_i = \left[x_i^1, x_i^2, \ldots, x_i^d\right]$ is a vector with d features, is used to train this ELM. Predictors in the input layer are connected to hidden nodes by randomly generated weights and

T. Lee et al., *Deep Learning for Hydrometeorology and Environmental Science*, Water Science and Technology Library 99,
https://doi.org/10.1007/978-3-030-64777-3_8

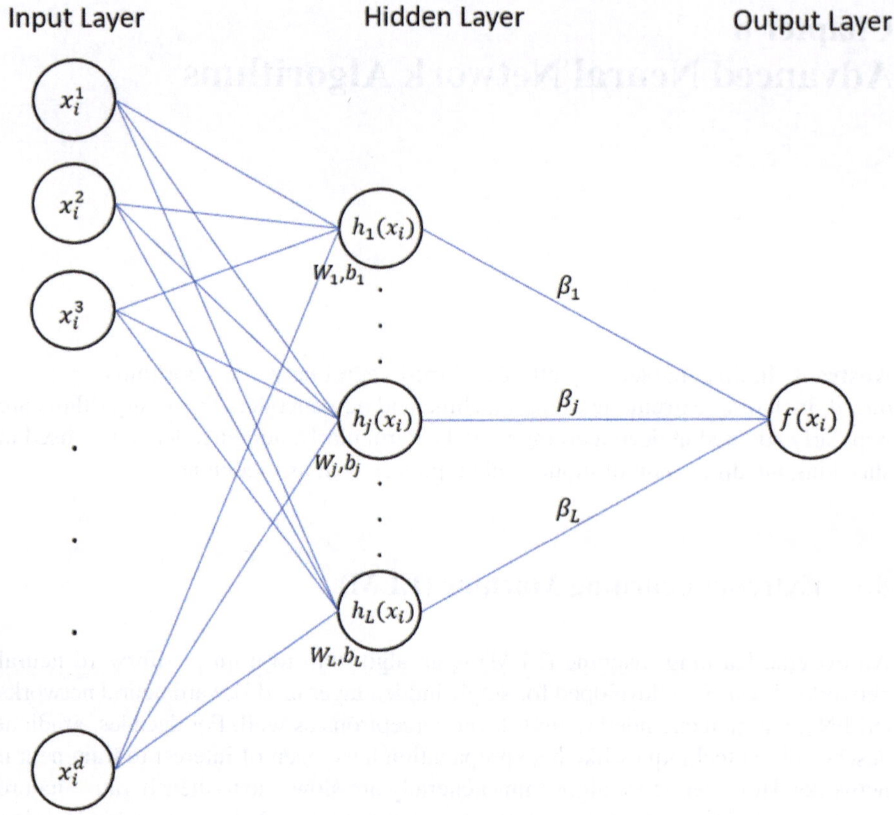

Fig. 8.1 Structure of a single layer feedforward neural network subject to training with ELM

bias $\{w_j, b_j\}_{j=1}^{L}$, called hidden node parameters. Although any continuous probability distribution can be used for the initialization of these parameters, the uniform distribution on the interval [−1,1] and the standard normal distribution have been commonly used in the literature.

In the hidden layer, nodes transform a d-dimensional input space to an L-dimensional ELM feature space. The output vector of the hidden layer, $h(x) = [h_1(x), h_2(x), \ldots, h_L(x)]^T$, is called the ELM nonlinear feature mapping. Each element of this vector, $h_j(x)$, can be written as

$$h_j(x) = G\big(w_j, b_j, x\big) \tag{8.1}$$

where the activation function $G(w, b, x)$ is a nonlinear piecewise continuous function satisfying ELM universal approximation capability theorems, such as tanh or sigmoid functions in Eqs. (4.3) and (4.5), respectively. Examples of commonly used activation

Table 8.1 Commonly used functions in ELM as activation function of hidden neurons

Name	Equations
Sigmoid	$G(w, b, x) = \frac{1}{1+\exp(-(w\cdot x+b))}$
Radial base	$G(w, b, x) = \exp(-(w \cdot x + b)^2)$
Hard-limit	$G(w, b, x) = \begin{cases} 1 \text{ if } w \cdot x + b \leq 0 \\ 0 \text{ otherwise} \end{cases}$
Sine	$G(w, b, x) = \sin(w \cdot x + b)$
Cosine	$G(w, b, x) = cos(w \cdot x + b)$
Triangle base	$G(w, b, x) = \begin{cases} 1 - \lvert w \cdot x + b\rvert & \text{if} \lvert a \cdot x + b\rvert < 1 \\ 0 & \text{otherwise} \end{cases}$
Multiquadric	$G(w, b, x) = (x - w + b^2)^2$
Hyperbolic tangent	$G(w, b, x) = \frac{1-\exp(-(w\cdot x+b))}{1+\exp(-(w\cdot x+b))}$

functions in ELM are presented in Table 8.1. Combinations of different types of activation may be used in ELM to achieve better performance.

The output function of ELM with L hidden nodes is

$$f_L(x_i) = \sum_{j=1}^{L} \beta_j h_j(x_i) = h(x_i)\beta \tag{8.2}$$

where $\beta = [\beta_1, \beta_2, \ldots, \beta_L]^T$ is the output weight vector connecting the L nodes of the hidden layer to the output node. These output weights are found through the learning process. Note unlike conventional neural networks, the ELM output function $f_L(x_i)$ does not need any bias.

In the following, we will discuss more about algorithms of basic ELM and generalized ELM to train the described SLFN using a training set with N distinct samples, each of which has d features and a target variable.

8.1.1 Basic ELM

The goal of basic ELM is to find the output weight vector (β) by minimizing the squared loss of the prediction error:

$$\min_{\beta \in \Re^{L\times 1}} \|\mathbf{H}\boldsymbol{\beta} - \mathbf{Y}\|^2 \tag{8.3}$$

where || || denotes the Frobenius norm, and **H** is the hidden layer output matrix calculated as:

$$\mathbf{H} = \begin{bmatrix} \mathbf{h}(x_1) \\ \vdots \\ \mathbf{h}(x_N) \end{bmatrix} = \begin{bmatrix} h_1(x_1) & \cdots & h_L(x_1) \\ \vdots & \vdots & \vdots \\ h_1(x_N) & \cdots & h_L(x_N) \end{bmatrix}_{N\times L}$$

$$= \begin{bmatrix} G(w_1 \cdot x_1 + b_1) & \cdots & G(w_L \cdot x_1 + b_L) \\ \vdots & \vdots & \vdots \\ G(w_1 \cdot x_N + b_1) & \cdots & G(w_L \cdot x_N + b_L) \end{bmatrix}_{N\times L} \quad (8.4)$$

The optimal solution for the ELM minimization problem can be obtained by:

$$\hat{\mathbf{b}} = \mathbf{H}^{*} \cdot \mathbf{Y} \quad (8.5)$$

where $\mathbf{H}^{*}$ is the Moore-Penrose generalized inverse of matrix which can be calculated by different methods, such as orthogonal projection method, orthogonalization method, iterative method, and singular value decomposition.

8.1.2 Generalized ELM

The generalized ELM attempts to find output weights by reaching not only the smallest training error but also the smallest norm of output weights. Therefore, the minimization problem in a generalized ELM can be written as:

$$\min_{\beta \in \Re^{L\times 1}} \|\boldsymbol{\beta}\|_p^{\sigma_1} + \mathrm{C}\|\mathbf{Y} - \mathbf{Hb}\|_q^{\sigma_2} \quad (8.6)$$

where $\sigma_1, \sigma_2 > 0$ and p, q $= 0, \frac{1}{2}, 1, 2, \ldots, +\infty$. The first term adds regularization ability to the objective function and controls the complexity of model. $C > 0$ is a tuning parameter selected through the validation process. C balances the regularization term and training error, therefore avoids overfitting and improves the stability and generalization performance of ELM. In a special case that $\sigma_1 = \sigma_2 = p = q = 2$, the above minimization problem is called ridge regression or regularized least squares:

$$\min_{\beta \in \Re^{L\times 1}} \frac{1}{2}\|\boldsymbol{\beta}\|^2 + \frac{C}{2}\|\mathbf{Y} - \mathbf{H}\boldsymbol{\beta}\|^2 \quad (8.7)$$

This minimization is a convex optimization problem which can be solved by setting the gradient of the regularized ELM objective function with respect to $\boldsymbol{\beta}$ to zero:

Table 8.2 Dataset of observed X, Y, and predicted Y as well as their errors for the example of ELM

i	x_i^1	x_i^2	$y_{o,i}$	$y_{p,i}$
1	0.51	0.69	0.58	0.2472
2	0.69	0.58	0.56	0.2480
3	0.58	0.56	0.61	0.2471
4	0.56	0.61	0.60	0.2472
5	0.61	0.60	0.64	0.2475
6	0.60	0.64	0.53	0.2476
7	0.64	0.53	0.47	0.2474
8	0.53	0.47	0.54	0.2463
9	0.47	0.54	0.44	0.2462
10	0.54	0.44	0.62	0.2461

$$\nabla_{\mathrm{ELM}} = \boldsymbol{\beta} - C\mathbf{H}^T(\mathbf{Y} - \mathbf{H}\boldsymbol{\beta}) = 0 \tag{8.8}$$

The ELM gradient equation can be easily solved analytically and the closed-form solution can be written in 2 forms, depending on N and L:

$$\hat{\mathbf{b}} = (\mathbf{H}^T\mathbf{H} + \frac{1}{C}\mathbf{I}_L)^{-1}\mathbf{H}^T\mathbf{Y} \quad \text{if N} > L \tag{8.9}$$

$$\hat{\mathbf{b}} = \mathbf{H}^T(\mathbf{H}\mathbf{H}^T + \frac{1}{C}\mathbf{I}_N)^{-1}\mathbf{Y} \quad \text{if N} < L \tag{8.10}$$

where **I** is an identity matrix.

Example 8.1 Train an SLFN by ELM. The network consists of two input nodes, one output node, and 20 hidden nodes. The training input (X) and output (Y) dataset are shown in Table 8.2 in the 2nd–4th columns. The initial weights and biases are shown in Table 8.3. Note that the initial weights and biases are randomly generated by uniform distributions, respectively, between [−1,1], and [0,1]. The predicted values with the initial set are shown in the 5th column of Table 8.2 as well as the error in the 6th column. Assume the penalty factor C is 0.1, and the number of hidden neurons is 20.

Solution:

At the next step, we build the input matrix $\boldsymbol{X}$ and target vector $\boldsymbol{Y}$. The feature matrix has 10 rows ($N = 10$) and two columns ($d = 2$):

$$\mathbf{X} = \begin{bmatrix} x_1 \\ x_2 \\ x_3 \\ \vdots \\ \vdots \\ x_{10} \end{bmatrix} = \begin{bmatrix} x_1^1 & x_1^2 \\ x_2^1 & x_2^2 \\ x_3^1 & x_3^2 \\ \vdots & \vdots \\ \vdots & \vdots \\ x_{10}^1 & x_{10}^2 \end{bmatrix} = \begin{bmatrix} 0.51 & 0.69 \\ 0.69 & 0.58 \\ 0.58 & 0.56 \\ \vdots & \vdots \\ \vdots & \vdots \\ 0.54 & 0.44 \end{bmatrix}$$

$$\mathbf{Y} = \begin{bmatrix} y_1 \\ y_2 \\ y_3 \\ \vdots \\ \vdots \\ y_{10} \end{bmatrix} = \begin{bmatrix} 0.58 \\ 0.56 \\ 0.61 \\ \vdots \\ \vdots \\ 0.62 \end{bmatrix}$$

The third step is to use the randomly generated weights (Table 8.3) to build matrix $\boldsymbol{H}$. This matrix has 10 rows ($N = 10$) and 10 columns ($L = 20$).

$$\mathrm{H} = \begin{bmatrix} \mathbf{h}(x_1) \\ \vdots \\ \mathbf{h}(x_{10}) \end{bmatrix} = \begin{bmatrix} h_1(x_1) & \cdots & h_{20}(x_1) \\ \vdots & \vdots & \vdots \\ h_1(x_{10}) & \cdots & h_{20}(x_{10}) \end{bmatrix}_{10\times 20}$$

$$= \begin{bmatrix} G(w_1 \cdot x_1 + b_1) & \cdots & G(w_{20} \cdot x_1 + b_{20}) \\ \vdots & \vdots & \vdots \\ G(w_1 \cdot x_{10} + b_1) & \cdots & G(w_{20} \cdot x_{10} + b_{20}) \end{bmatrix}_{10\times 20}$$

In this example, the sigmoid function $G(w, b, x) = \frac{1}{1+\exp(-(w.x+b))}$ is used as the activation function of hidden neurons:

$$\mathbf{H} = \begin{bmatrix} G\left(\mathrm{w}_1^1 * \mathrm{x}_1^1 + \mathrm{w}_1^2 * \mathrm{x}_1^2 + b_1\right) & \cdots & G\left(\mathrm{w}_{20}^1 * \mathrm{x}_1^1 + \mathrm{w}_{20}^2 * \mathrm{x}_1^2 + b_{20}\right) \\ \vdots & \vdots & \vdots \\ G\left(\mathrm{w}_1^1 * \mathrm{x}_{10}^1 + \mathrm{w}_1^2 * \mathrm{x}_{10}^2 + b_1\right) & \cdots & G\left(\mathrm{w}_{20}^1 * \mathrm{x}_{10}^1 + \mathrm{w}_{20}^2 * \mathrm{x}_{10}^2 + b_{20}\right) \end{bmatrix}_{10\times 20}$$

$$= \begin{bmatrix} G(-0.731 * 0.51 - 0.322 * 0.69 + 0.111) & \cdots & G(0.353 * 0.51 + 0.756 * 0.69 + 0.471) \\ \vdots & \vdots & \vdots \\ G(-0.731 * 0.54 - 0.322 * 0.44 + 0.111) & \cdots & G(0.353 * 0.54 + 0.756 * 0.44 + 0.471) \end{bmatrix}$$

$$= \begin{bmatrix} G(-0.484) & \cdots & G(1.172) \\ \vdots & \vdots & \vdots \\ G(-0.425) & \cdots & G(0.994) \end{bmatrix}_{10\times 20} = \begin{bmatrix} \frac{1}{1+\exp(0.484)} & \cdots & \frac{1}{1+\exp(-1.172)} \\ \vdots & \vdots & \vdots \\ \frac{1}{1+\exp(0.425)} & \cdots & \frac{1}{1+\exp(0.994)} \end{bmatrix}_{10\times 20}$$

$$= \begin{bmatrix} 0.381 & \cdots & 0.764 \\ \vdots & \vdots & \vdots \\ 0.395 & \cdots & 0.730 \end{bmatrix}_{10\times 20}$$

Since L = 20 > N = 10: $\hat{\boldsymbol{\beta}} = \mathbf{H}^T(\mathbf{H}\mathbf{H}^T + \frac{1}{C}\mathbf{I}_N)^{-1}\mathbf{Y}$

$$\mathbf{H}^T = \begin{bmatrix} 0.381 & \cdots & 0.395 \\ \vdots & \vdots & \vdots \\ 0.764 & \cdots & 0.730 \end{bmatrix}_{20*10}$$

$$\mathbf{H}\mathbf{H}^T = \begin{bmatrix} 0.381 & \cdots & 0.764 \\ \vdots & \vdots & \vdots \\ 0.395 & \cdots & 0.730 \end{bmatrix}_{10\times 20} * \begin{bmatrix} 0.381 & \cdots & 0.395 \\ \vdots & \vdots & \vdots \\ 0.764 & \cdots & 0.730 \end{bmatrix}_{20*10} = \begin{bmatrix} 7.934 & \cdots & 7.887 \\ \vdots & \vdots & \vdots \\ 7.887 & \cdots & 7.868 \end{bmatrix}_{10*10}$$

$$\frac{1}{C}\mathbf{I}_N = \frac{1}{0.01}\begin{bmatrix} 1 & \cdots & 0 \\ \vdots & \vdots & \vdots \\ 0 & \cdots & 1 \end{bmatrix}_{10\times 10} = \begin{bmatrix} 100 & \cdots & 0 \\ \vdots & \vdots & \vdots \\ 0 & \cdots & 100 \end{bmatrix}_{10\times 10}$$

$$\left(\mathbf{H}\mathbf{H}^T + \frac{1}{C}\mathbf{I}_N\right)^{-1} = \left(\begin{bmatrix} 107.934 & \cdots & 7.887 \\ \vdots & \vdots & \vdots \\ 7.887 & \cdots & 107.868 \end{bmatrix}_{10*10}\right)^{-1} = \begin{bmatrix} 0.0095 & \cdots & -0.0004 \\ \vdots & \vdots & \vdots \\ -0.0004 & \cdots & 0.0095 \end{bmatrix}_{10*10}$$

$$\hat{\boldsymbol{\beta}} = \begin{bmatrix} 0.381 & \cdots & 0.395 \\ \vdots & \vdots & \vdots \\ 0.764 & \cdots & 0.730 \end{bmatrix}_{20*10} * \begin{bmatrix} 0.0095 & \cdots & -0.0004 \\ \vdots & \vdots & \vdots \\ -0.0004 & \cdots & 0.0095 \end{bmatrix}_{10*10} * \begin{bmatrix} 0.58 \\ 0.56 \\ 0.61 \\ \vdots \\ \vdots \\ 0.62 \end{bmatrix}_{10*1}$$

$$= \begin{bmatrix} 0.0021 & \cdots & 0.0023 \\ \vdots & \vdots & \vdots \\ 0.0043 & \cdots & 0.0040 \end{bmatrix}_{20*10} * \begin{bmatrix} 0.58 \\ 0.56 \\ 0.61 \\ \vdots \\ \vdots \\ 0.62 \end{bmatrix} = \begin{bmatrix} 0.012 \\ 0.021 \\ 0.015 \\ \vdots \\ \vdots \\ 0.023 \end{bmatrix}$$

The output weights can be found in Table 8.2. Finally, the predicted values can be calculated as:

$$\mathbf{Y}_{\text{Prediction}} = \mathbf{H}\boldsymbol{\beta}$$

$$\mathbf{Y}_{\text{Prediction}} = \begin{bmatrix} 0.381 & \cdots & 0.764 \\ \vdots & \vdots & \vdots \\ 0.395 & \cdots & 0.730 \end{bmatrix}_{10\times 20} * \begin{bmatrix} 0.012 \\ 0.021 \\ 0.015 \\ \vdots \\ \vdots \\ 0.023 \end{bmatrix}_{20\times 1} = \begin{bmatrix} 0.2472 \\ 0.2480 \\ 0.2471 \\ \vdots \\ \vdots \\ 0.2461 \end{bmatrix}_{10\times 1}$$

Table 8.3 Initial weights and biases for the first layer of ELM

j	w_j^1	w_j^2	b_j	β_j
1	−0.731	−0.322	0.111	0.012
2	0.226	−0.563	0.920	0.021
3	−0.541	−0.572	0.605	0.015
4	0.698	−0.775	0.537	0.019
5	0.872	−0.572	0.900	0.023
6	0.295	0.791	0.529	0.024
7	−0.123	0.897	0.134	0.020
8	0.810	0.895	0.718	0.026
9	0.846	−0.882	0.317	0.018
10	−0.694	0.561	0.728	0.020
11	−0.517	−0.983	0.202	0.011
12	0.370	0.717	0.303	0.022
13	0.287	−0.326	0.326	0.018
14	−0.363	0.835	0.622	0.022
15	0.948	−0.395	0.310	0.020
16	−0.435	0.397	0.519	0.019
17	−0.395	0.522	0.368	0.019
18	−0.284	0.213	0.123	0.016
19	−0.517	−0.685	0.479	0.014
20	0.353	0.756	0.471	0.023

Comparing the predicted value and observed value shows that the network with assumed hyper parameters (number of hidden neurons and penalty factor) was not quite successful to predict the true values of y. To achieve the best possible performance, all the hyper parameters should be selected through a validation process.

8.1.3 Python Coding

Here, the python coding for Example 8.1 is presented as follows.

```
[Python 8.1] Python code for ELM in Example 8.1
import numpy as np
from numpy.linalg import inv
from sklearn.metrics import mean_squared_error

def train_ELM(x,y,n_hidden,activation,C) :
    n_feature = x.shape[1]    #Number of input variables
    n_sample = x.shape[0]     #Number of samples in training set
    bias = np.random.uniform(1,0,n_hidden).reshape(1,n_hidden)
    bias_matrix = np.repeat(bias,n_sample, axis=0)
    input_weight = np.random.uniform(-1,1,n_hidden*n_feature).reshape(n_hidden,n_feature)
    tempH = np.matmul(x,input_weight.transpose())
    tempH = tempH + bias_matrix
   # activation : 'sigmoid':
      H = 1 / (1 + np.exp(-1 * tempH))
```

```
      I = np.identity(n_hidden)
      output_weight = np.matmul(np.matmul(inv(np.matmul(H.transpose(),H) + I),H.transpose()),y)
    else:
      I = np.identity(n_sample)
      output_weight = np.matmul(np.matmul(H.transpose(),inv(np.matmul(H,H.transpose())+I)),y)
    y_pred = np.matmul(H, output_weight)
    return(bias,input_weight,output_weight, y_pred)
```

```
#Set the hyper parameters
n_hidden_neuron = 20
C = 0.01
activation = 'sigmoid'

#Reading training input and output
X_data = np.array([[0.51, 0.69],[0.69, 0.58],[0.58, 0.56],[0.56, 0.61],[0.61, 0.6 ], [0.6 , 0.64], [0.64, 0.53],
                   [0.53, 0.47], [0.47, 0.54], [0.54, 0.44]])
Y_data = np.array([[0.58],[0.56],[0.61],[0.60],[0.64],[0.53],[0.47],[0.54],[0.44],[0.62]])


#Call train_ELM function for given training set
B,W,beta,prediction = train_ELM(X_data, Y_data, n_hidden, activation,C)
```

8.2 Autoencoder

An autoencoder (AE) is an unsupervised neural network that aims to reconstruct its input variables. In this group of neural networks, the target values in the output layer are equal to the input variables, and AE is trained to generate an approximation of inputs as accurately as possible. Each autoencoder consists of two key parts, as shown in Fig. 8.2:

- Encoder: seeks to map input variables into a latent space representation, which may have a higher or lower dimension than the original input space, depending on the type of autoencoder. If the dimension of encoded representation is less

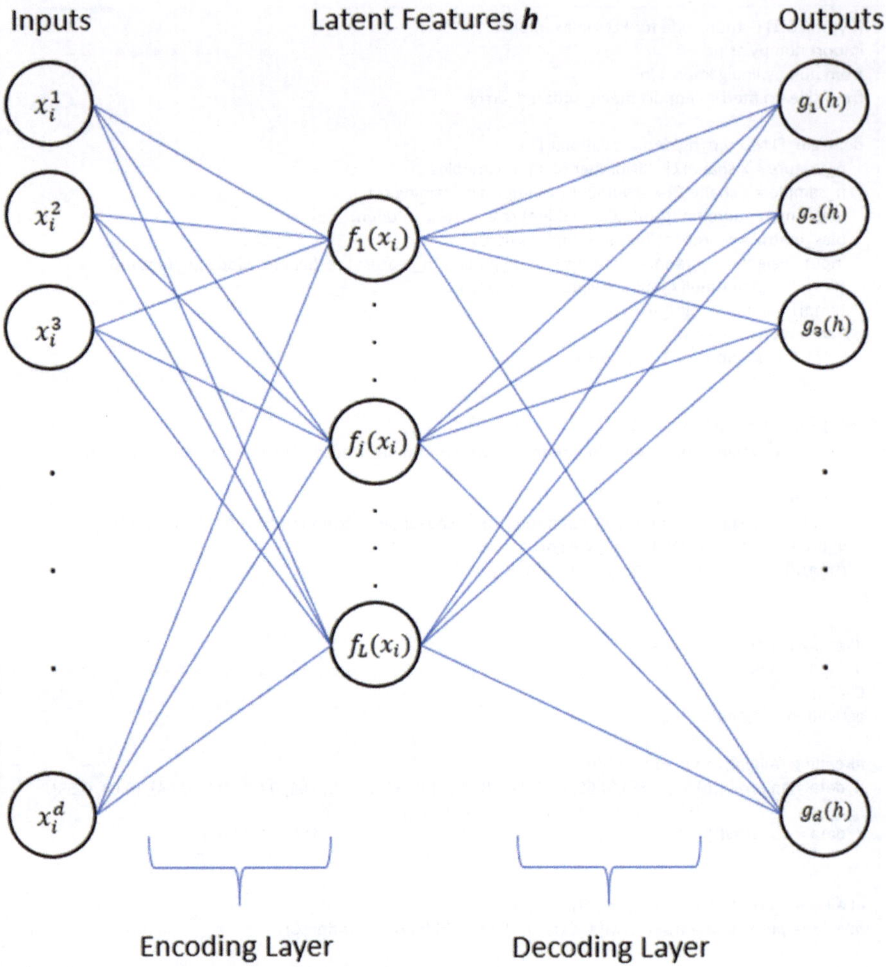

Fig. 8.2 Structure of an autoencoder with a single hidden layer

than the dimension of input, the network is called under complete autoencoder. Otherwise, it is named overcomplete autoencoder. An under complete autoencoder can be considered similar to PCA but with more capacity and flexibility. Both can generate a subspace from input data. However, autoencoder, unlike PCA, is able to use linear and non-linear encoding functions, which make it more flexible.

- Decoder: uses encoded values to produce a copy of input variables.

What usually makes autoencoder interesting and beneficial is not the decoder output but the latent representation. It is able to present the original data in a more efficient or a more detailed form, depending on the type of autoencoder. The hidden representation of an under complete autoencoder is a compressed form of the original input data, while it contains all the important information about it. Therefore, it can be used in high dimensional problems instead of original data to overcome the curse of dimensionality. The application of latent features is not limited to dimension reduction. Sometimes more information is needed beyond what the original representation of data provides. In this situation, the latent representation of an overcomplete AE can help since this type of autoencoder expands on the input space and reveals some new information about input data.

8.2.1 *Vanilla Autoencoder*

The simplest form of an autoencoder is a single hidden layer neural network which learns how to regenerate its input (Fig. 8.2). The sizes of the input layer and the output layer are similar. Depending on the application of the encoded layer, the size of hidden neurons in the code layer can be smaller or larger than that of the other layers. The encoder function ($\mathbf{h} = f(\mathbf{X})$) and decoder function ($g(\mathbf{h})$) can be any nonlinear function. Like other neural networks, the best number of hidden neurons and the most appropriate functions for encoder and decoder are the learning parameters that should be selected based on the minimum validation error.

If the mean squared error is selected as a loss function, the learning process of this simple autoencoder aims to minimize $(\mathbf{X} - g(f(\mathbf{X}))^2$. The choice of optimizer depends on the computational cost and time that one is willing to pay.

Example 8.2 Build a simple undercomplete AE with three hidden neurons using the training and validation data sets provided in Tables 8.4 and 8.5. Assume a learning rate = 0.5, epochs = 50, and n_hidden = 3. Use the sigmoid function (f(x), g(h)) for all neurons in the hidden layer and the output layer.

For this example, a simple feed forward neural network with one hidden layer is trained. The variables in the training set are used as input. The network must learn a way to regenerate these variables in output which means $y_{o,i} = X_{o,i}$

To train this network, the loss function is defined as:

$$E = \frac{1}{2}\sum_{i=1}^{N}(x_{o,i} - x_{p,i})^2 = \frac{1}{2}\sum_{i=1}^{N} e_i^2$$

Backpropagation and gradient descent algorithms with constant learning rate are used here to build a vanilla autoencoder.

First, initialize the network. Use seed = 6 to get the same random numbers as this example. Then, feed the network and calculate the latent representation and prediction for each sample (X_i) as follows:

$$\mathbf{X}_{\mathrm{Latent_i}} = \mathbf{h}_i = f(\mathbf{W}_{IH}\mathbf{X}_i + \mathbf{b})$$

$$= f\left\{\begin{bmatrix} w_{1,1}^{IH} & \cdots & w_{6,1}^{IH} \\ \vdots & \vdots & \vdots \\ w_{1,3}^{IH} & \cdots & w_{6,3}^{IH} \end{bmatrix}_{3\times 6} * \begin{bmatrix} x_i^1 \\ \vdots \\ x_i^6 \end{bmatrix} + \mathbf{b}\right\}$$

For example, for the first sample in the training set, the latent representation for the first epoch can be written as:

$$\mathbf{h}_1 = f\left\{\begin{bmatrix} 0.89 & \cdots & 0.59 \\ \vdots & \vdots & \vdots \\ 0.64 & \cdots & 0.82 \end{bmatrix}_{3\times 6} * \begin{bmatrix} 0.53 \\ \vdots \\ 0.44 \end{bmatrix} + \begin{bmatrix} 0.53 \\ \vdots \\ 0.05 \end{bmatrix}_{3*1}\right\}$$

$$= \begin{bmatrix} f(1.92) \\ \vdots \\ f(2.4) \end{bmatrix}_{3*1} = \begin{bmatrix} \frac{1}{1+\exp(-1.92)} \\ \vdots \\ \frac{1}{1+\exp(-2.4)} \end{bmatrix}_{3*1} = \begin{bmatrix} 0.87 \\ \vdots \\ 0.92 \end{bmatrix}_{3*1}$$

$$\mathbf{X}_{\text{prediction}_i} = g(\mathbf{W}_{HO}\mathbf{h}_i + \mathbf{b})$$

$$\mathbf{X}_{\text{prediction}_1} = g\left\{\begin{bmatrix} 0.72 & \cdots & 0.74 \\ \vdots & \vdots & \vdots \\ 0.73 & \cdots & 0.99 \end{bmatrix}_{6\times 3} * \begin{bmatrix} 0.87 \\ \vdots \\ 0.92 \end{bmatrix} + \begin{bmatrix} 0.71 \\ \vdots \\ 0.45 \end{bmatrix}_{6*1}\right\}$$

$$\mathbf{X}_{\text{prediction}_1} = \begin{bmatrix} g(0.94) \\ \vdots \\ g(0.66) \end{bmatrix}_{6*1} = \begin{bmatrix} \frac{1}{1+\exp(-0.94)} \\ \vdots \\ \frac{1}{1+\exp(-0.66)} \end{bmatrix}_{6*1} = \begin{bmatrix} 0.72 \\ \vdots \\ 0.66 \end{bmatrix}_{6*1}$$

The next step is back propagation and updating the weights (see Chap. 5). As an example, the calculations for $w_{1,1}^{HO}$ which is the weight associated with the connection between the first neuron in the encoded layer (hidden layer) and the one in the output layer are presented here:

$$w_{1,1}^{HO} = w_{1,1}^{HO} - \text{learning rate} * \frac{\partial E}{\partial w_{1,1}^{HO}} = 0.72 + 0.5 * (0.53 - 0.28) * \left(\frac{0.28}{1 - 0.28}\right) * 0.13$$

$$w_{1,1}^{HO} = 0.73$$

Continue updating the weights for all the samples, 50 times (epochs = 50).

After training, the mean square error of this simple autoencoder over the validation dataset would be equal to 0.0251.

Table 8.4 Dataset of observed X for the example of AE

	i	x_i^1	x_i^2	x_i^3	x_i^4	x_i^5	x_i^6
Training data	1	0.53	0.47	0.51	0.89	0.45	0.44
	2	0.51	0.69	0.49	0.77	0.38	0.37
	3	0.64	0.53	0.60	0.86	0.52	0.50
	4	0.60	0.64	0.56	0.80	0.45	0.44
	5	0.58	0.56	0.55	0.85	0.46	0.45
	6	0.61	0.60	0.57	0.83	0.47	0.46
	7	0.56	0.61	0.53	0.82	0.44	0.43
	8	0.54	0.44	0.51	0.90	0.47	0.45
Validation data	1	0.47	0.54	0.45	0.86	0.39	0.38
	2	0.69	0.58	0.64	0.84	0.53	0.51

Table 8.5 Dataset of noisy X for the example of AE

	i	$\tilde{x}_i^1$	$\tilde{x}_i^2$	$\tilde{x}_i^3$	$\tilde{x}_i^4$	$\tilde{x}_i^5$	$\tilde{x}_i^6$
Training data	1	0.69	0.34	0.47	1.14	0.33	0.50
	2	0.27	0.63	0.26	0.94	0.30	0.29
	3	1.05	0.71	0.18	0.95	0.56	0.73
	4	0.83	0.42	0.53	0.60	0.57	0.51
	5	0.50	0.82	0.80	0.38	0.47	0.52
	6	0.69	0.47	0.75	0.26	0.67	0.53
	7	0.68	0.37	0.54	1.18	0.68	0.24
	8	0.52	0.39	0.59	0.63	0.23	0.28
Validation data	1	0.29	0.79	0.71	0.89	0.22	0.31
	2	1.02	0.52	0.55	1.38	0.63	0.54

Python coding of Vanilla AE

The python coding for Example 8.2 is presented as follows.

```
[Python 8.2] Python code for Vanilla AE in Example. 8.2

import numpy as np
from sklearn.model_selection import train_test_split

#Initialize the network
def initialize_network(n_inputs, n_hidden, n_outputs):
    network = list()
    encoded_layer = [{'weights':[np.random.random() for i in range(n_inputs + 1)]} for i in range(n_hidden)]
    network.append(encoded_layer)
    output_layer = [{'weights':[np.random.random() for i in range(n_hidden + 1)]} for i in range(n_outputs)]
    network.append(output_layer)
    return network
```

```
# Calculate input to each neuron
def Hidden_input(weights, inputs):
    hidden_input_value = 0
    for i in range(len(weights)-1):
        hidden_input_value += weights[i] * inputs[i]
    hidden_input_value += weights[-1] # add bias
    return hidden_input_value

# Transfer with sigmoid function
def Hidden_transfer(hidden_input_value):
    return 1.0 / (1.0 + np.exp(-hidden_input_value))
```

```
# Feed Forward Step
def feed_forward(network, row):
        inputs = row
        for layer in network:
                new_inputs = []
                for neuron in layer:
                        hidden_input_value = Hidden_input(neuron['weights'], inputs)
                        neuron['output'] = Hidden_transfer(hidden_input_value)
                        new_inputs.append(neuron['output'])
                inputs = new_inputs
        return inputs
```

[Python 1.2] Continued

```
#calculate the derivative of a neuron
def sigmoid_derivative(output):
        return output * (1.0 - output)


# Backpropagattion Step
def backward_propagate_error(network, observation,n_row,epoch):
        for i in reversed(range(len(network))):
                layer = network[i]
                errors = list()
                if i != len(network)-1:
                        for j in range(len(layer)):
                                error = 0.0
                                for neuron in network[i + 1]:
                                        error += (neuron['weights'][j] * neuron['delta'])
```

```
                else:
                        for j in range(len(layer)):
                                neuron = layer[j]
                                errors.append(observation[j] - neuron['output'])
                for j in range(len(layer)):
                        neuron = layer[j]
                        neuron['delta'] = errors[j] * sigmoid_derivative(neuron['output'])
```

```
# Update network weights with constant learning rate
def update_weights(network, row, l_rate):
  for i in range(len(network)):
    inputs = row
    if i != 0:
      inputs = [neuron['output'] for neuron in network[i - 1]]
    for neuron in network[i]:
      for j in range(len(inputs)):
        neuron['weights'][j] += l_rate * neuron['delta'] * inputs[j]

        neuron['weights'][-1] += l_rate * neuron['delta']
```

```
[Python 1.2] Continued

# Train the network
def train_network(network, x_train, l_rate, n_epoch, n_outputs):
  for epoch in range(n_epoch):
    sum_error = 0
    n_row=0
    for row in x_train:
      prediction = feed_forward(network, row)
      observation = row
      sum_error += sum([(prediction[i]-observation[i])**2 for i in range(len(observation))])
      backward_propagate_error(network, observation,n_row,epoch)
      update_weights(network, row, l_rate)
      n_row= n_row+1
    print('>epoch=%d, lrate=%.3f, error=%.3f' % (epoch, l_rate, sum_error))
```

```
encoding_dim = 3 #Number of neurons in the hidden layer
n_i = 6  #Number of input variables
n_o = 6 #Number of neurons in the output layer

#Reading training input and output
X_train = np.array ([[0.53 , 0.47, 0.51, 0.89, 0.45 , 0.44],  [0.51 , 0.69 , 0.49, 0.77, 0.38,0.37],
     [0.64, 0.53, 0.60, 0.86, 0.52,0.50], [0.6,0.64 ,0.56, 0.80, 0.45,0.44], [0.58 ,0.5, 0.55, 0.85, 0.46,0.45],
     [0.61, 0.6, 0.57, 0.83, 0.47,0.46], [0.56 ,0.61,0.53, 0.82, 0.44,0.43], [0.54, 0.44, 0.51, 0.90, 0.47, 0.45]])

X_val = np.array ([[0.47,0.54,0.45,0.86,0.39,0.38],[0.69,0.58,0.64,0.84,0.53,0.51]])
```

```
# Train AE
seed = 6
network = initialize_network(n_i, encoding_dim, n_o)
output_test = feed_forward(network, X_train[0])
train_network(network, X_train, 0.5, 50, n_outputs)

#Calculate the validation error
sum_error = 0
for row in X_val:
  prediction = feed_forward(network, row)
  sum_error += sum([(prediction[i]-row[i])**2 for i in range(len(row))])
MSE_validation = sum_error / len(X_val)
```

8.2.2 Regularized Autoencoder

Although a simple autoencoder may learn some features about input distribution, it can easily learn a way to just memorize and copy the training data without extracting valuable information from the data. This situation is similar to what we call over-fitting in other learning techniques, and it happens when the architecture of autoencoder and capabilities of encoding and decoding functions do not match the complexity of input data. To avoid this problem, regularized autoencoders have been developed in which their learning process aims to improve their performance not only by minimizing the reconstruction errors but by controlling some of their other properties such as their sparsity or robustness. There are various forms of regularization that can be employed in the network in order to encourage good generalization properties. Here we only talk about sparse autoencoders and denoising autoencoders.

Sparse Autoencoder

In a sparse AE, the code is constrained to have sparsity; otherwise, the activation functions within this layer will be penalized in proportion to the magnitude of their activation. This training strategy forces the AE to activate a small number of neurons in the hidden layer and simultaneously minimize the error. Therefore, to achieve the least possible reconstruction error, the network has to keep only those neurons activated, which extract the most salient features of input data.

One should note the difference between activity or sparsity regularization and weight regularization. A sparsity regularizer imposes a penalty on the activity of the latent representations rather than weights and as a result, it controls the number of units that would fire (or be activated) at a given time.

There are two ways to impose sparsity constraints: L1 regularization and KL-Divergence. Here, only the application of an L1 activity regularizer is discussed. The minimization problem of an $L1$ regularized AE can be written as:

$$\text{Min Loss}(\mathbf{X}, \text{g}(\text{f}(\mathbf{X}))) + \lambda |f(\mathbf{X})| \tag{8.11}$$

The first term represents the reconstruction error, which shows how much $g(f(X))$ differs from X. The choice of a loss function depends on the problem that needs to be dealt with. It can be a cross-entropy loss function in a classification problem, and it can be represented by the mean square error function in a regression problem. The second term is called activity regularizer, and it penalizes the $L1$ norm of the output vector of the hidden layer (or activations). The penalty term λ is a hyperparameter that defines the amount or degree that the $L1$ norm will be weighted and penalized. The fully trained sparse AE has the capability to keep the neurons inactive as much as possible.

The penalty coefficient λ should be found through a validation process. Common values are on a logarithmic scale between 0 and 0.1. One can start the search from course values like 0.1 and then continue the search on a fine logarithmic grid until the minimum validation error is achieved.

Denoising Autoencoder

Denoising Autoencoder (DAE) adopts a different approach towards regulating the network rather than penalizing the latent representation. This form of regularization takes on a noisy and corrupted copy of X ($\tilde{X}$) as input and aims to regenerate the original form of data. In a denoising AE, the training process attempts to minimize $\mathrm{Loss}\left(\mathbf{X}, \mathrm{g}\left(\mathrm{f}\left(\tilde{\mathbf{X}}\right)\right)\right)$. Therefore, unlike the previous AEs, the input and output of DAE are no longer identical and the network must learn generalizable encoder and decoder, which are able to discover the most interesting and meaningful aspects of data structure among noises. Adding random noise to x and feeding AE with this new representation of data prevents AE from simply copying the input to output without learning its important features.

Example 8.3 Build a simple, an L1 regularized and a denoising undercomplete AE using the training and validation sets provided in Tables 8.4 and 8.5. In this example, the "keras" package of Python was used to build all models. Keras is a powerful package that enables us to implement deep learning algorithms very easily.

1. Vanilla AE

Two dense layers with the sigmoid activation function were used for this AE. The encoded layer had three hidden neurons, representing the latent features, and the output layer had six neurons representing six regenerated input variables. Normal distribution was used for initializing the weights for both layers (kernel_initializer = "normal").

After building the model, the loss function and optimization method should be selected. Here, the mean square error and Adam optimizer were used. In the next step, the model was fitted using training X. Since we were looking for training an AE, the target variable was the same as input. Because the size of our data set was very small and the simple AE does not use any kind of generalization, this model was very prone to get over-fitted. Therefore, it was trained only on 50 epochs. The batch size was set to 2, and validation data were used to calculate the validation error. The Python code are presented in what follows. After training, the validation error was calculated. The error for this model was 0.0068.

2. L1 Sparse AE

For this AE, a hidden layer with three hidden neurons and sigmoid function were employed as encoder layer. An L1 activity regularizer were added to this layer ($\lambda = 0.00001$) in order to generalize the performance of this model. At decoder layer, six neurons with sigmoid activation function were used to regenerate the six input variables. Both layers were initialized randomly using the normal distribution.

The same as Vanilla AE, the mean square error and Adam optimizer were used for this L1 Sparse AE. Since this model takes advantage of L1 regularization, it may be reasonable to use more epochs without facing the risk of over-fitting. The validation error calculated for this model was 0.0061.

3 Denoising AE

To train this type of AE, a noisy and corrupted form of variables is needed. Random noises with

The standard normal distribution (mean = 0, standard deviation = 1) was added to the original data and used as input variables to train AE. The noisy data is shown in Table 8.5.

The same as the previous AEs, three and six neurons with sigmoid function were used in the hidden layer and output layer, respectively. Note that the noisy data and original data were, respectively, introduced to the model as input and output data. The validation error (MSE) of the trained denoising model was 0.0057.

Note that in real applications, one needs to select an appropriate activation function, number of neurons, number of layers, number of epochs, number of batch sizes, and … through the validation process to make sure that the best possible performance is achieved.

8.2.3 Python Coding of Regularized AE

The python coding for Example 8.3 is presented as follows.

```
[[Python 8.2] Continued
] Python code for Vanilla AE in Example. 8.3

import numpy as np
from keras.models import Sequential
from keras.layers import Dense
from keras import regularizers

encoding_dim = 3 #Number of neurons in hidden layer
n_i = 6  #Number of input variables
```

```
#Reading training input and output
X_train = np.array ([[0.53 , 0.47, 0.51, 0.89, 0.45 , 0.44], [0.51 , 0.69 , 0.49, 0.77, 0.38,0.37],
      [0.64, 0.53, 0.60, 0.86, 0.52,0.50], [0.6,0.64 ,0.56, 0.80, 0.45,0.44], [0.58 ,0.5, 0.55, 0.85, 0.46,0.45],
      [0.61, 0.6, 0.57, 0.83, 0.47,0.46], [0.56 ,0.61,0.53, 0.82, 0.44,0.43], [0.54, 0.44, 0.51, 0.90, 0.47, 0.45]])

X_val = np.array ([[0.47,0.54,0.45,0.86,0.39,0.38],[0.69,0.58,0.64,0.84,0.53,0.51]])

X_train_noise = np.array ([[0.69,0.34,0.47,1.14,0.33,0.50], [0.27,0.63,0.26,0.94,0.30,0.29],
[1.05,0.71,0.18,0.95,0.56,0.73], [0.83,0.42,0.53,0.60,0.57,0.51], [0.50,0.82,0.80,0.38,0.47,0.52],
[0.69,0.47,0.75,0.26,0.67,0.53], [0.68, 0.37, 0.54,1.18,0.68,0.24], [0.52,0.39,0.59,0.63,0.23,0.28]])

X_val_noise = np.array ([[0.29,0.79,0.71,0.89,0.22,0.31],[1.02,0.52,0.55,1.38,0.63,0.54]])

#Build the simplest AE
model = Sequential()
model.add (Dense ( encoding_dim, input_dim=n_i, kernel_initializer='normal', activation='sigmoid' ))
model.add (Dense ( n_i, kernel_initializer='normal', activation='sigmoid' ))

#Fit the model
model.compile ( loss='mean_squared_error', optimizer='adam' )
history = model.fit ( X_train, X_train, epochs=50, batch_size=2 ,validation_data=(X_val,X_val),verbose=0 )
```

[Python 8.3] Python code for L1 regularized and denoising AE in Example. 8.3

```
#Build the L1 regularized AE AE
model_sparse = Sequential()
model_sparse.add( Dense(encoding_dim, input_dim=n_i, activation='sigmoid', kernel_initializer='normal',
                activity_regularizer=regularizers.l1(0.00001)))
model_sparse.add( Dense(n_i, kernel_initializer='normal', activation='sigmoid'))

#Fit the model
model_sparse.compile ( loss='mean_squared_error', optimizer='adam' )
history_sparse = model_sparse.fit ( X_train, X_train, epochs=100,
                                batch_size=2 ,validation_data=(X_val,X_val),verbose=0)

#Build a Denoising AE
model_denoise = Sequential()
model_denoise.add( Dense( encoding_dim, input_dim=n_i, kernel_initializer='normal',
                activation='sigmoid' ))
model_denoise.add( Dense( n_i, kernel_initializer='normal', activation='sigmoid' ))

#Fit the model
model_denoise.compile ( loss='mean_squared_error', optimizer='adam' )
history_denoise = model_denoise.fit ( X_train_noise, X_train, epochs=100,
                                batch_size=2 ,validation_data=(X_val_noise,X_val),verbose=0)
```

8.3 Exercises

1. Construct an SLFN by ELM for the dataset in Table 2.1. Note that the network consists of 2 input nodes, 1 output node, and 10 hidden nodes. The training input (X) and output (Y) dataset are shown in Table 2.1 in the 2nd–4th columns.
   (a) Train the SLFN by ELM. Assume the penalty factor and C as 0.2.
   (b) Estimate the Y and plot the estimate and observed value.
   (c) Train the SLFN by ELM for the penalty factors as $C = 0.2$, 0.5, and 0.8 and discuss how the training and estimates are differed.
2. Program the python code to run exercise 1 above employing train_ELM in the example of Python 8.1.
3. Build an under complete AE with two hidden neurons using the training and validation data sets provided in Table 8.4. Assume a learning rate = 0.5, epochs = 5. Use the sigmoid functions for $f(\mathbf{X})$ and $g(h)$) for all neurons in the hidden layer and the output layer.

Reference

Huang GB, Zhu QY, Siew CK (2006) Extreme learning machine: theory and applications. Neurocomputing 70(1–3):489–501. https://doi.org/10.1016/j.neucom.2005.12.126

Chapter 9
Deep Learning for Time Series

Abstract One of the major applications in deep learning models is to forecast the future. In recent years, time series forecasting with deep learning models has been developed and applied in a number of fields. Recurrent neural network models can allow forecasting future better, and long short-term memory (LSTM) is a breakthrough to overcome the shortages of the previous RNN model. These algorithms are explained in detail in this chapter.

9.1 Recurrent Neural Network

Recurrent neural network (RNN) is one type of ANNs that the output from a previous step can be fed as input to the next step for a time series. This structure allows to have a memory capturing information about what has been performed up to the current step. RNNs contain a feedback loop so that the current hidden state serves as an additional input to estimate the subsequent hidden state. For RNNs, the hidden state vector can be modified from the traditional NN as:

$$\boldsymbol{h}_t = g_{IH}(\boldsymbol{W}_{IH}\boldsymbol{x}_t + \boldsymbol{W}_{HP}\boldsymbol{h}_{t-1} + b_{IH}) \tag{9.1}$$

where $\boldsymbol{W}_{HP}$ is the weight parameter matrix for the previous hidden state. To estimate the parameter set composed of weights and biases, backpropagation through time (BPTT), which is similar to BP in traditional NNs, is employed.

Recurrent networks use their understanding of past events to process the input vector. However, RNNs have a strong weakness, called gradient vanishing problem, such that the gradients backpropagating to the hidden units shrink exponentially when small. "tanh" is more often employed to overcome this problem for the activation function of the hidden state g_{IH} rather than sigmoid due to its gradient in the derivative. As shown in Fig. 4.7 the derivative of the sigmoid function shows a maximum of about 0.25 while the derivative of tanh function has a maximum of 1.0. The vanishing problem is worsened because a number of multiplications must be taken to the derivative of an activation function. However, the vanishing problem

T. Lee et al., *Deep Learning for Hydrometeorology and Environmental Science*, Water Science and Technology Library 99,
https://doi.org/10.1007/978-3-030-64777-3_9

still exists even with the tanh activation function due to many multiplications. The new type of RNN, called LSTM, has been introduced to solve the vanishing problem, introduced in the next section.

In the following, the backpropagation of the RNN is presented. A general backpropagation method cannot handle the memory of the previous hidden state. Subsequently, a new algorithm is introduced to implement the memory of the previous state.

9.1.1 Backpropagation

To illustrate the learning procedure of an RNN, the simple example of RNN is applied as shown in Fig. 9.1, similar to the simple neural network in Fig. 5.1 with the parameters of α_1 and w_1 as well as the one additional parameter u_1 for the previous condition. Its prediction can be made as

$$y_{p,t} = \alpha_1 h_t(s_t) = \alpha_1 h_t(w_1 x_t + u_1 h_{t-1}) \tag{9.2}$$

where $s_t = w_1 x_t + u_1 h_{t-1}$. Note that the tanh function is commonly employed for $h(.)$ to avoid the vanishing gradient problem in the sigmoid function.

To estimate the RNN parameters with the backpropagation approach, the gradient is estimated as:

For the parameter of α_1,

$$\frac{\partial E}{\partial \alpha_1} = \frac{\partial E}{\partial y_p}\frac{\partial y_p}{\partial \alpha_1} = -\sum_{t=1}^{N} e_t h_t \tag{9.3}$$

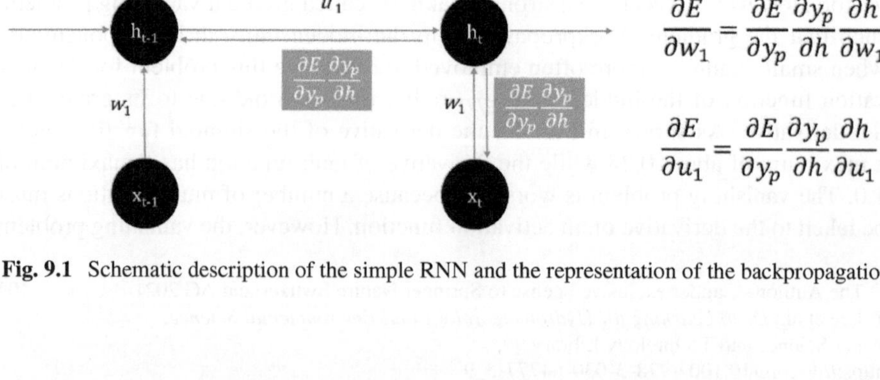

Fig. 9.1 Schematic description of the simple RNN and the representation of the backpropagation

For w_1,

$$\frac{\partial E}{\partial w_1} = \frac{\partial E}{\partial y_p}\frac{\partial y_p}{\partial h}\frac{\partial h}{\partial w_1} = -\sum_{t=1}^{N} e_t\alpha_1\left(1 - h_t^2\right)x_t \tag{9.4}$$

where

$$\frac{\partial y_p}{\partial h} = \frac{\partial(\alpha_1 h)}{\partial h} = \alpha_1$$

$$\frac{\partial h}{\partial w_1} = \frac{\partial h}{\partial s}\frac{\partial s}{\partial u_1} = \left(1 - h_t^2\right)\frac{\partial s}{\partial u_1} = \left(1 - h_t^2\right)x_t$$

$$\frac{\partial h(s)}{\partial s} = 1 - h_t^2$$

Since the tanh function is commonly employed for $h(.)$.
To update the weight for the previous hidden state u_1,

$$\frac{\partial E}{\partial u_1} = \frac{\partial E}{\partial y_p}\frac{\partial y_p}{\partial u_1} = \frac{\partial E}{\partial y_p}\frac{\partial y_p}{\partial h}\frac{\partial h}{\partial u_1} = -\sum_{t=1}^{N} e_t\alpha_1\left(1 - h_t^2\right)h_{t-1} \tag{9.5}$$

where

$$\frac{\partial h}{\partial u_1} = \frac{\partial h}{\partial s}\frac{\partial s}{\partial u_1} = \left(1 - h_t^2\right)\frac{\partial s}{\partial u_1} = \left(1 - h_t^2\right)h_{t-1}$$

9.1.2 Backpropagation Through Time (BPTT)

The backpropagation method is not sufficient to update the weights. Therefore, backpropagation through time (BPTT) has been suggested as below:

$$\frac{\partial E}{\partial \boldsymbol{w}} = \sum_t \frac{\partial E_t}{\partial \boldsymbol{w}} = \sum_t \frac{\partial E_t}{\partial y_{p,t}}\frac{\partial y_{p,t}}{\partial h_t}\frac{\partial h_t}{\partial \boldsymbol{w}} \tag{9.6}$$

Here, $\boldsymbol{w}$ indicates the parameter set that is in the function of h_t such as u_1 and w_1 in Eq. (9.2). The gradient of the other parameter α_1 can be easily estimated as

$$\frac{\partial E}{\partial \alpha_1} = \frac{\partial E}{\partial y_p}\frac{\partial y_p}{\partial \alpha_1} = -\sum_{t=1}^{N} e_t h_t \tag{9.7}$$

Since h_t is the function of h_{t-1} and so on, the term $\frac{\partial E_t}{\partial w}$ can be expressed as

$$\frac{\partial E_t}{\partial w} = \frac{\partial E_t}{\partial y_{p,t}} \frac{\partial y_{p,t}}{\partial h_t} \frac{\partial h_t}{\partial w} = \sum_{k=1}^{t} \frac{\partial E_t}{\partial y_{p,t}} \frac{\partial y_{p,t}}{\partial h_t} \frac{\partial h_t}{\partial h_{t-k}} \frac{\partial h_{t-k}}{\partial w} \tag{9.8}$$

where

$$\frac{\partial h_t}{\partial w} = \frac{\partial h_t}{\partial h_{t-1}} \frac{\partial h_{t-1}}{\partial w} + \cdots + \frac{\partial h_t}{\partial h_1} \frac{\partial h_1}{\partial w} = \sum_{k=1}^{t} \frac{\partial h_t}{\partial h_{t-k}} \frac{\partial h_{t-k}}{\partial w} \tag{9.9}$$

and

$$\frac{\partial h_t}{\partial h_{t-k}} = \frac{\partial h_t}{\partial h_{t-1}} \frac{\partial h_{t-1}}{\partial h_{t-2}} \cdots \frac{\partial h_{t-k+1}}{\partial h_{t-k}} = \prod_{j=1}^{k} \frac{\partial h_{t-j+1}}{\partial h_{t-j}} \tag{9.10}$$

$$\frac{\partial E_t}{\partial w} = \frac{\partial E_t}{\partial y_{p,t}} \frac{\partial y_{p,t}}{\partial h_t} \frac{\partial h_t}{\partial w} = \sum_{k=1}^{t} \frac{\partial E_t}{\partial y_{p,t}} \frac{\partial y_{p,t}}{\partial h_t} \left(\prod_{j=1}^{k} \frac{\partial h_{t-j+1}}{\partial h_{t-j}} \right) \frac{\partial h_{t-k}}{\partial w} \tag{9.11}$$

When this is applied to u_1,

$$\frac{\partial E_t}{\partial u_1} = \sum_{k=1}^{t} \frac{\partial E_t}{\partial y_{p,t}} \frac{\partial y_{p,t}}{\partial h_t} \left(\prod_{j=1}^{k} \frac{\partial h_{t-j+1}}{\partial h_{t-j}} \right) \frac{\partial h_{t-k}}{\partial u_1} \tag{9.12}$$

Since

$$\frac{\partial h_{t-k}}{\partial u_1} = \left[1 - \tanh(s_{t-k})^2\right] h_{t-k-1} \tag{9.13}$$

$$\frac{\partial h_{t-j+1}}{\partial h_{t-j}} = \left[1 - \tanh\left(s_{t-j+1}\right)^2\right] u_1 \tag{9.14}$$

Therefore,

$$\begin{aligned} \frac{\partial E_t}{\partial u_1} &= \frac{\partial E_t}{\partial y_{p,t}} \frac{\partial y_{p,t}}{\partial h_t} \frac{\partial h_t}{\partial u_1} \\ &= -\sum_{k=1}^{t} e_t \alpha_1 \left(\prod_{j=1}^{k} \left[1 - \tanh\left(s_{t-j+1}\right)^2\right] u_1 \right) \left[1 - \tanh(s_{t-k})^2\right] h_{t-k-1} \end{aligned} \tag{9.15}$$

Finally,

$$\frac{\partial E}{\partial u_1} = \sum_t \frac{\partial E_t}{\partial u_1}$$
$$= -\sum_t \sum_{k=1}^{t} e_t \alpha_1 \left(\prod_{j=1}^{k} \left[1 - \tanh\left(s_{t-j+1}\right)^2 \right] u_1 \right) \left[1 - \tanh(s_{t-k})^2 \right] h_{t-k-1} \tag{9.16}$$

For w_1,

$$\frac{\partial h_{t-k}}{\partial w_1} = \left[1 - \tanh(s_{t-k})^2 \right] x_t \tag{9.17}$$

Therefore,

$$\frac{\partial E}{\partial w_1} = \sum_t \frac{\partial E_t}{\partial u_1}$$
$$= -\sum_t \sum_{k=1}^{t} e_t \alpha_1 \left(\prod_{j=1}^{k} \left[1 - \tanh\left(s_{t-j+1}\right)^2 \right] u_1 \right) \left[1 - \tanh(s_{t-k})^2 \right] x_{t-k} \tag{9.18}$$

For example, $t = 3$:

$$\frac{\partial E_3}{\partial w} = \frac{\partial E_3}{\partial y_{p,t}} \frac{\partial y_{p,3}}{\partial h_3} \left\{ \frac{\partial h_{3-1+1}}{\partial h_{3-1}} \frac{\partial h_{3-1}}{\partial w} + \frac{\partial h_{3-1+1}}{\partial h_{3-1}} \frac{\partial h_{3-2+1}}{\partial h_{3-2}} \frac{\partial h_{3-2}}{\partial w} + \frac{\partial h_{3-1+1}}{\partial h_{3-1}} \frac{\partial h_{3-2+1}}{\partial h_{3-2}} \frac{\partial h_{3-3+1}}{\partial h_{3-3}} \frac{\partial h_{3-3}}{\partial w} \right\}$$
$$= \frac{\partial E_3}{\partial y_{p,t}} \frac{\partial y_{p,3}}{\partial h_3} \left\{ \frac{\partial h_3}{\partial h_2} \frac{\partial h_2}{\partial w} + \frac{\partial h_3}{\partial h_2} \frac{\partial h_2}{\partial h_1} \frac{\partial h_1}{\partial w} + \frac{\partial h_3}{\partial h_2} \frac{\partial h_2}{\partial h_1} \frac{\partial h_1}{\partial h_0} \frac{\partial h_0}{\partial w} \right\}$$

$$\frac{\partial E_3}{\partial u_1} = -e_3 \alpha_1 \{ (1 - \tanh(s_3)^2) u_1 [1 - \tanh(s_2)^2] h_1 + (1 - \tanh(s_3)^2)$$
$$u_1 (1 - \tanh(s_2)^2) u_1 [1 - \tanh(s_1)^2] h_0 + (1 - \tanh(s_3)^2) u_1 (1 - \tanh(s_2)^2)$$
$$u_1 (1 - \tanh(s_1)^2) u_1 [1 - \tanh(s_0)^2] h_{-1} \}$$

where let $h_j = 0$ at $j \leq 0$.

$$\frac{\partial E_3}{\partial w_1} = -e_3 \alpha_1 \left\{ \left(1 - \tanh(s_3)^2\right) u_1 \left[1 - \tanh(s_2)^2\right] x_2 \mathrm{S} + \left(1 - \tanh(s_3)^2\right) u_1 \left(1 - \tanh(s_2)^2\right) \right.$$
$$u_1 \left[1 - \tanh(s_1)^2\right] x_1 + \left(1 - \tanh(s_3)^2\right) u_1 \left(1 - \tanh(s_2)^2\right)$$
$$\left. u_1 \left(1 - \tanh(s_1)^2\right) u_1 \left[1 - \tanh(s_0)^2\right] x_0 \right\}$$

Example 9.1 Update the weights of the RNN (i.e. u_1, w_1 and α_1) in Fig. 9.1 with the BPTT method. Here, the current u_1, w_1, and α_1 are 0.7, 0.8 and −0.3, respectively, and the learning rate is 0.01 and employ the dataset as in Table 5.1.

Solution:

For example,

(1) $y_{p,i}$ is calculated with (9.2) and $e_i = y_{o,i} - y_{p,i}$. For example, t = 3:

$$y_{p,3} = -0.3h_t(0.8 \times 0.34 + 0.7 \times 0.63) = -0.3 \times 0.61 = -0.18$$
$$e_3 = 0.2 - (-0.18) = 0.38$$

(2) Estimate $\frac{\partial E}{\partial \alpha_1}$, $\frac{\partial E}{\partial w_1}$, and $\frac{\partial E}{\partial u_1}$ with BPTT

$$\frac{\partial E}{\partial \alpha_1} = -\sum_t eh_t = -1.043$$

For calculating $\frac{\partial E}{\partial u_1} = \sum_t \frac{\partial E_t}{\partial u_1}$
or example, t = 3:

$$\frac{\partial E_3}{\partial u_1} = -e_3\alpha_1\{(1 - \tanh(s_3)^2)u_1[1 - \tanh(s_2)^2]h_1\}$$
$$= -0.38(-0.3)(1 - 0.61^2)0.7(1 - 0.63^2)0.14 = 0.0042$$

$$\frac{\partial E}{\partial u_1} = -0.2889$$
$$\frac{\partial E}{\partial w_1} = -0.0366$$

(3) Then,

$$\begin{bmatrix} w_1 \\ u_1 \\ \alpha_1 \end{bmatrix}^{(2)} = \begin{bmatrix} w_1 \\ u_1 \\ \alpha_1 \end{bmatrix}^{(1)} - \eta \begin{bmatrix} \frac{\partial E}{\partial w_1} \\ \frac{\partial E}{\partial u_1} \\ \frac{\partial E}{\partial \alpha_1} \end{bmatrix} = \begin{bmatrix} 0.8 \\ 0.7 \\ -0.3 \end{bmatrix} - 0.01 \begin{bmatrix} -0.0366 \\ -0.2889 \\ -1.043 \end{bmatrix} = \begin{bmatrix} 0.8004 \\ 0.7029 \\ -0.2896 \end{bmatrix}$$

(4) Repeat steps (1)–(3).
(5) Repeat step (4).

```
[Python 1.1] Program the backpropagation through time (BPTT) algorithm learning
def bptt(Ypred,Yobs,X,W,A,U,ht,lr):
    T=len(X)
    delta_ht=np.zeros(T)
    d_ObsPre=Yobs-Ypred
    for t in np.arange(T)[::-1]:
        delta_ht[t]=dtanH(ht[t])*U

    dEdU1=np.zeros(T)
    dEdW1=np.zeros(T)
    htp=np.insert(ht,0,1.0)
    bp_lookback=10
```

```
    for t in np.arange(T)[::-1]:
        dEdh=A*(d_ObsPre[t])
        s1_U=np.zeros(t+1)
        s1_W=np.zeros(t+1)
        nStart=max(0,t-bp_lookback)
        for k in np.arange(nStart,t)[::-1]:
            s1_U[k]=dEdh*np.prod(delta_ht[k:t])*htp[k]*dtanH(ht[k])
            s1_W[k]=dEdh*np.prod(delta_ht[k:t])*X[k]*dtanH(ht[k])
```

```
        dEdU1[t]=np.sum(s1_U)
        dEdW1[t]=np.sum(s1_W)

    dEdU=-np.sum(dEdU1)
    dEdW=-np.sum(dEdW1)
    dEdA=-np.sum((d_ObsPre)*ht)

    W -= lr*dEdW
    A -= lr*dEdA
    U -= lr*dEdU
    return W, A, U
```

9.2 Long Short-Term Memory (LSTM)

9.2.1 Basics of LSTM

Hochreiter and Schmidhuber (1997) proposed a particular RNN model, called Long Short-Term Memory (LSTM) to overcome the vanishing problem in RNNs. In this LSTM model, multiplicative gate units learn to open and close the access to the information flow.

Three gates (as input, output, and forget gates) in LSTM control the amount of information flow within the cell that is responsible for memorizing values over a number of time intervals, as shown in Fig. 9.2b. Three gates calculate an activation of a weighted sum with the sigmoid function ranging from zero to one as described below:

$$\mathbf{f}_t = \sigma\left(\mathbf{W}_{fx}\mathbf{x}_t + \mathbf{W}_{fh}\mathbf{h}_{t-1} + b_f\right) \tag{9.19}$$

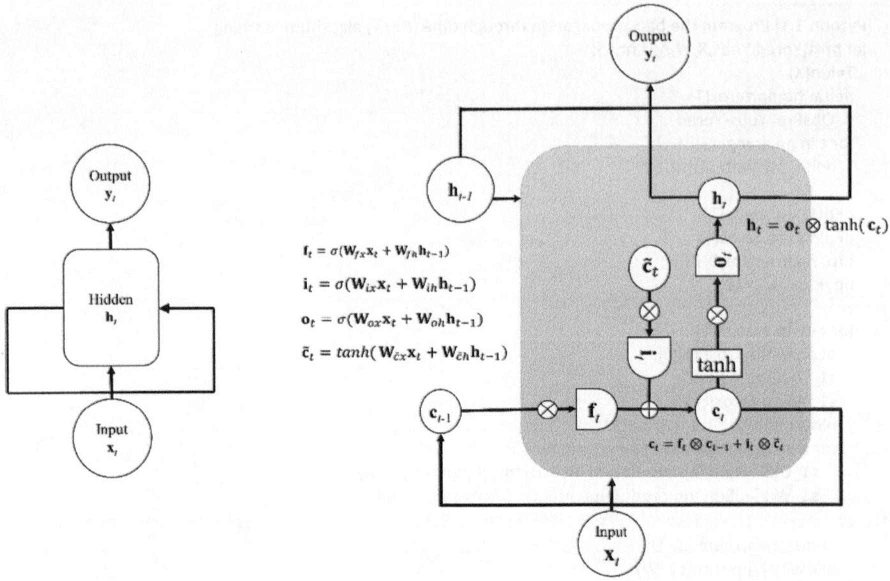

(a) RNN (b) LSTM

Fig. 9.2 Diagram of RNN and LSTM

$$\boldsymbol{i}_t = \sigma(\mathbf{W}_{ix}\mathbf{x}_t + \mathbf{W}_{ih}\mathbf{h}_{t-1} + b_i) \tag{9.20}$$

$$\boldsymbol{o}_t = \sigma(\mathbf{W}_{ox}\mathbf{x}_t + \mathbf{W}_{oh}\mathbf{h}_{t-1} + b_o) \tag{9.21}$$

where $\sigma(z)$ indicates the sigmoid function in Eq. (4.3), and $\mathbf{W}$ and b are the weight parameter matrix and bias. The candidate cell state ($\tilde{\boldsymbol{c}}_t$) is defined with the tanh activation function ranging as $[-1, 1]$ as:

$$\tilde{\mathbf{c}}_t = \tanh(\mathbf{W}_{\tilde{c}x}\mathbf{x}_t + \mathbf{W}_{\tilde{c}h}\mathbf{h}_{t-1} + b_{\tilde{c}}) \tag{9.22}$$

The cell state can be described with two components as the amount of the previous cell state information and the magnitude of the current updated information denoted as

$$\mathbf{c}_t = \mathbf{f}_t \otimes \mathbf{c}_{t-1} + \mathbf{i}_t \otimes \tilde{\mathbf{c}}_t \tag{9.23}$$

where $\otimes$ represents the element multiplication. The magnitude of information to be flown in is decided by the gates here: forget gate ($\mathbf{f}_t$) and input gate ($\mathbf{i}_t$). For example, if the forget gate is zero, then the previous cell information is lost and only the current new cell information is updated, that is, $\mathbf{c}_t = \mathbf{i}_t \otimes \tilde{\mathbf{c}}_t$. Eventually, the hidden state is defined by how much the current cell state information goes out to the hidden state as:

$$\mathbf{h}_t = \mathbf{o}_t \otimes \tanh(\mathbf{c}_t) \tag{9.24}$$

The diagram shown in Fig. 9.2b presents that output $\mathbf{y}_t$ is a linear combination of hidden state $\mathbf{h}_t$. The hidden state is the cell state filtered with the *tanh* function whose intensity is controlled with the output gate $\mathbf{o}_t$. The cell state is a linear combination of the current updated cell information ($\tilde{\boldsymbol{c}}_t$) and the information inflow from the previous cell state ($\mathbf{c}_t$).

This deep learning algorithm can sequentially deliver long-term memory through the cell state. As stated earlier, if the forget gate is zero, then the previous cell information is lost, and only the current new cell information is updated, that is, $\boldsymbol{c}_t = \boldsymbol{i}_t \otimes \tilde{\boldsymbol{c}}_t$. Conversely, the previous cell state information is completely delivered as $\boldsymbol{c}_t = \boldsymbol{c}_{t-1}$ when $\mathbf{i}_t = 0$ and $\mathbf{f}_t = 1$. Note that those three gates are defined with the previous hidden state ($\mathbf{h}_{t-1}$) and the current input ($\mathbf{x}_t$, here y_{t-1}). The reason that this algorithm has the long-term memory ability is due to the particular information delivery system with the cell state and three control gates in addition to the hidden state.

The final output is generally made up of the linear function of the hidden state as

$$\mathbf{y}_t = \mathbf{W}_{hy}\mathbf{h}_t + b_{hy} \tag{9.25}$$

9.2.2 *Example of LSTM*

One example of LSTM prediction with two hidden units and one input and one output variable is presented with the estimated parameter set for the weights and three gates (Table 9.1). Referring to the LSTM structure to Fig. 9.2b, the previous

Table 9.1 Example of LSTM parameter set with 2 hidden units

		$\mathbf{W}_x$	$\mathbf{W}_h$		b
Hidden 1	$\mathbf{i}_t$	0.293	0.191	−0.574	0.864
	$\mathbf{f}_t$	0.164	−0.449	0.463	0.978
	$\tilde{\boldsymbol{c}}_t$	0.806	−0.386	0.355	0.234
	$\mathbf{o}_t$	0.419	0.556	−0.655	0.872
Hidden 2	$\mathbf{i}_t$	0.253	0.029	−0.332	0.766
	$\mathbf{f}_t$	0.338	−0.408	0.393	1.088
	$\tilde{\boldsymbol{c}}_t$	−0.699	0.220	−0.150	−0.306
	$\mathbf{o}_t$	0.579	0.518	−0.588	0.782

cell states ($\mathbf{c}_{t-1} = [0.587, -1.146]$) and hidden states ($\mathbf{h}_{t-1} = [0.454, -0.696]$) are shown in Table 9.2 as well as the output layer setup. The current input value is $x_t = 0.269$, as shown in Table 9.3.

From the input value and the hidden states, the values of the input, forget, and output gates are estimated, as shown in Table 9.3. For example, the input gate values $\boldsymbol{i}_t = [i_t^1; i_t^2]$ are estimated as:

$$i_t^1 = \sigma(\boldsymbol{W}_{ix}\boldsymbol{x}_t + \boldsymbol{W}_{ih}\boldsymbol{h}_{t-1} + b_i)$$

$$= \sigma\left(0.293 \times 0.269 + [0.191, -0.574]\begin{bmatrix} 0.454 \\ -0.696 \end{bmatrix} + 0.864\right)$$

$$= \sigma(1.429) = 1/(1 + \exp(-1.429) = 0.807$$

$$i_t^2 = \sigma\left(0.253 \times 0.269 + [0.029, -0.332]\begin{bmatrix} 0.454 \\ -0.696 \end{bmatrix} + 0.766\right)$$

$$= \sigma(1.088) = 0.746$$

The candidate cell state $\tilde{\boldsymbol{c}}_t$ is defined as

$$\tilde{c}_t^1 = \tanh(\boldsymbol{W}_{\tilde{c}x}\boldsymbol{x}_t + \boldsymbol{W}_{\tilde{c}h}\boldsymbol{h}_{t-1} + b_{\tilde{c}}) = \tanh(0.806 \times 0.269$$

Table 9.2 Example of the LSTM parameter and dataset. Note that the bold numbers show the estimated value in this example

	t−1	t (updated)	NN	
$\mathbf{c}_t$	0.587	**0.388**	w_1	0.723
	−1.146	**−0.981**	w_2	−0.710
$\mathbf{h}_t$	0.454	**0.312**	B	−0.256
	−0.696	**0.076**	Y	**−0.084**
x_t		0.269		

Table 9.3 Example of the estimated gates and cell state updates

		St	**g(St)**
Hidden 1	$\mathbf{i}_t$	1.429	0.807
	$\mathbf{f}_t$	0.496	0.622
	$\tilde{c}_t$	0.029	0.029
	$\mathbf{o}_t$	1.693	0.845
Hidden 2	$\mathbf{i}_t$	1.078	0.746
	$\mathbf{f}_t$	0.720	0.673
	$\tilde{c}_t$	−0.289	−0.282
	$\mathbf{o}_t$	1.583	0.830

$$+ \ [-0.386, 0.355]\begin{bmatrix} 0.454 \\ -0.696 \end{bmatrix} + 0.234) = \tanh(0.029) = 0.029$$

$$\tilde{c}_t^2 = \tanh(-0.289) = -0.282$$

With these gate values, the cell state and hidden state are updated:

$$\boldsymbol{c}_t = \begin{bmatrix} c_t^1 \\ c_t^2 \end{bmatrix} = \boldsymbol{f}_t \otimes \boldsymbol{c}_{t-1} + \boldsymbol{i}_t \otimes \tilde{\boldsymbol{c}}_t$$

$$= \begin{bmatrix} 0.622 \\ 0.673 \end{bmatrix} \otimes \begin{bmatrix} 0.587 \\ -1.146 \end{bmatrix} + \begin{bmatrix} 0.807 \\ 0.746 \end{bmatrix} \otimes \begin{bmatrix} 0.029 \\ -0.282 \end{bmatrix} = \begin{bmatrix} 0.388 \\ -0.981 \end{bmatrix}$$

$$\boldsymbol{h}_t = \begin{bmatrix} h_t^1 \\ h_t^2 \end{bmatrix} = \boldsymbol{o}_t \otimes \tanh(\boldsymbol{c}_t)$$

$$= \begin{bmatrix} 0.845 \\ 0.830 \end{bmatrix} \otimes \begin{bmatrix} \tanh(0.388) \\ \tanh(-0.981) \end{bmatrix} = \begin{bmatrix} 0.312 \\ 0.076 \end{bmatrix}$$

Finally, the predicted value can be defined, with the weight and bias parameter set shown on the right side of Table 9.2, as

$$y_t = \boldsymbol{W}_{ho}\boldsymbol{h}_t + b_{ho} = [0.723 - 0.710]\begin{bmatrix} 0.312 \\ 0.076 \end{bmatrix} - 0.256 = -0.084$$

Note that a simple linear activation function is employed as an output activation function.

9.2.3 Backpropagation of a Simple LSTM

Here, a simple LSTM is introduced to illustrate the backpropagation algorithm to update the weights of the LSTM. The system has only one hidden node and only one input without a bias term (i.e., b). The three gates (forget, input, and output) are denoted as:

$$f_t = \sigma\left(W_{fx}x_t + W_{fh}h_{t-1}\right) \tag{9.26}$$

$$i_t = \sigma(W_{ix}x_t + W_{ih}h_{t-1}) \tag{9.27}$$

$$o_t = \sigma(W_{ox}x_t + W_{oh}h_{t-1}) \tag{9.28}$$

The cell candidate state and the cell state are defined as

$$\tilde{c}_t = \tanh(W_{\tilde{c}x} x_t + W_{\tilde{c}h} h_{t-1}) \tag{9.29}$$

$$c_t = f_t \otimes c_{t-1} + i_t \otimes \tilde{c}_t \tag{9.30}$$

The hidden state for this system is

$$h_t = o_t \otimes \tanh(c_t) \tag{9.31}$$

Finally, the output is defined with a linear activation function as:

$$y_t = W_{hy} h_t \tag{9.32}$$

By explicitly describing the simple LSTM network,

$$y_t = W_{hy}[o_t \otimes \tanh(c_t)] \tag{9.33}$$

$$y_t = W_{hy}\left[o_t \otimes (f_t \otimes c_{t-1} + i_t \otimes \tilde{c}_t)\right] \tag{9.34}$$

$$\begin{aligned} y_t = W_{hy}\big[&\sigma(W_{ox} x_t + W_{oh} h_{t-1}) \otimes \tanh\{\sigma\left(W_{fx} x_t + W_{fh} h_{t-1}\right) \otimes c_{t-1} \\ &+ \sigma(W_{ix} x_t + W_{ih} h_{t-1})\{\otimes \tanh(W_{\tilde{c}x} x_t + W_{\tilde{c}h} h_{t-1})\}\big] \end{aligned} \tag{9.35}$$

For the backward propagation,

$$\frac{\partial E}{\partial y_t} = -\sum_{t=1}^{N} e_i = -\sum_{t=1}^{N} y_{o,t} - y_{p,t} \tag{9.36}$$

From, $y_t = W_{hy} h_t$

$$\frac{\partial E}{\partial W_{hy}} = \frac{\partial E}{\partial y_p} \frac{\partial y_p}{\partial W_{hy}} == -\sum_{t=1}^{N} \left(y_{o,t} - y_{p,t}\right) h_t \tag{9.37}$$

Also,

$$\begin{aligned} y_{p,t} &= W_{hy} h_t = W_{hy}[o_t \otimes \tanh(c_t)] \\ &= W_{hy}[\sigma(W_{ox} x_t + W_{oh} h_{t-1}) \otimes \tanh(c_t)] \end{aligned} \tag{9.38}$$

$$\begin{aligned} \frac{\partial E}{\partial W_{ox}} &= \frac{\partial E}{\partial y_{p,t}} \frac{\partial y_{p,t}}{\partial h_t} \frac{\partial h_t}{\partial o_t} \frac{\partial o_t}{\partial S_o} \frac{\partial S_o}{\partial W_{ox}} \\ &= -\sum_{t=1}^{N} \left(y_{o,t} - y_{p,t}\right) W_{hy} \tanh(c_t) d\sigma(S_o) x_t \end{aligned} \tag{9.39}$$

where $\frac{\partial h_t}{\partial o_t} = \tanh(c_t)$, $\frac{\partial o_t}{\partial S_o} = d\sigma(S_o)$ and $S_o = W_{ox}x_t + W_{oh}h_{t-1}$

$$\frac{\partial E}{\partial W_{oh}} = \frac{\partial E}{\partial y_{p,t}} \frac{\partial y_{p,t}}{\partial h_t} \frac{\partial h_t}{\partial o_t} \frac{\partial o_t}{\partial S_o} \frac{\partial S_o}{\partial W_{oh}}$$
$$= -\sum_{t=1}^{N} (y_{o,t} - y_{p,t}) W_{hy} \tanh(c_t) d\sigma(S_o) h_{t-1} \tag{9.40}$$

For W_{fx} and W_{fh}

$$\begin{aligned} y_t &= W_{hy}h_t = W_{hy}\left[o_t \otimes \mathrm{C^h}\right] = W_{hy}[o_t \otimes \tanh(c_t)] \\ &= W_{hy}\left[o_t \otimes (f_t \otimes c_{t-1} + i_t \otimes \tilde{c}_t)\right] \\ &= W_{hy}\left[o_t \otimes \left\{\sigma\left(W_{fx}x_t + W_{fh}h_{t-1}\right) \otimes c_{t-1} + i_t \otimes \tanh(\tilde{c}_t)\right\}\right] \end{aligned} \tag{9.41}$$

$$\frac{\partial E}{\partial W_{fh}} = \frac{\partial E}{\partial y_{p,t}} \frac{\partial y_{p,t}}{\partial h_t} \frac{\partial \mathrm{h_t}}{\partial \mathrm{C^h}} \frac{\partial \mathrm{C^h}}{\partial c_t} \frac{\partial c_t}{\partial f_t} \frac{\partial f_t}{\partial S_f} \frac{\partial S_f}{\partial W_{fh}}$$
$$= -\sum_{t=1}^{N} (y_{o,t} - y_{p,t}) W_{hy} o_t \ \mathrm{dtanh}(c_t) c_{t-1} d\sigma\left(S_f\right) h_{t-1} \tag{9.42}$$

$$\frac{\partial E}{\partial W_{fx}} = \frac{\partial E}{\partial y_{p,t}} \frac{\partial y_{p,t}}{\partial h_t} \frac{\partial \mathrm{h_t}}{\partial \mathrm{C^h}} \frac{\partial \mathrm{C^h}}{\partial c_t} \frac{\partial c_t}{\partial f_t} \frac{\partial f_t}{\partial S_f} \frac{\partial S_f}{\partial W_{fx}}$$
$$= -\sum_{t=1}^{N} (y_{o,t} - y_{p,t}) W_{hy} o_t \ \mathrm{dtanh}(c_t) c_{t-1} d\sigma\left(S_f\right) x_t \tag{9.43}$$

where $S_f = W_{fx}x_t + W_{fh}h_{t-1}$.

For W_{ix} and W_{ih}

$$\begin{aligned} y_t &= W_{hy}h_t = W_{hy}\left[o_t \otimes \mathrm{C^h}\right] = W_{hy}[o_t \otimes \tanh(c_t) \\ &= W_{hy}\left[o_t \otimes \tanh\{f_t \otimes c_{t-1} + i_t \otimes \tilde{c}_t\}\right] \\ &= W_{hy}\left[o_t \otimes \tanh\{f_t \otimes c_{t-1} + \sigma(W_{ix}x_t + W_{ih}h_{t-1} + b_i) \otimes \tilde{c}_t\}\right] \end{aligned} \tag{9.44}$$

$$\frac{\partial E}{\partial W_{ih}} = \frac{\partial E}{\partial y_{p,t}} \frac{\partial y_{p,t}}{\partial h_t} \frac{\partial \mathrm{h_t}}{\partial \mathrm{C^h}} \frac{\partial \mathrm{C^h}}{\partial c_t} \frac{\partial c_t}{\partial i_t} \frac{\partial i_t}{\partial S_i} \frac{\partial S_i}{\partial W_{ih}}$$
$$= -\sum_{t=1}^{N} (y_{o,t} - y_{p,t}) W_{hy} o_t \ \mathrm{dtanh}(c_t) \tilde{c}_t d\sigma(S_i) h_{t-1} \tag{9.45}$$

$$\frac{\partial E}{\partial W_{ix}} = \frac{\partial E}{\partial y_{p,t}} \frac{\partial y_{p,t}}{\partial h_t} \frac{\partial \mathrm{h_t}}{\partial \mathrm{C^h}} \frac{\partial \mathrm{C^h}}{\partial c_t} \frac{\partial c_t}{\partial i_t} \frac{\partial i_t}{\partial S_i} \frac{\partial S_i}{\partial W_{ix}}$$

$$= -\sum_{t=1}^{N}\left(y_{o,t} - y_{p,t}\right)W_{hy}o_t\ \text{dtanh}(c_t)\tilde{c}_t d\sigma(S_i)x_t \tag{9.46}$$

where $S_i = W_{ix}x_t + W_{ih}h_{t-1}$.

For $W_{\tilde{c}x}$ and $W_{\tilde{c}h}$,

$$\begin{aligned} y_t &= W_{hy}h_t = W_{hy}\left[o_t \otimes \mathrm{C^h}\right] = W_{hy}[o_t \otimes \tanh(c_t)] \\ &= W_{hy}\left[o_t \otimes \tanh\{f_t \otimes c_{t-1} + i_t \otimes \tilde{c}_t\}\right] \\ &= W_{hy}\left[o_t \otimes \tanh\{f_t \otimes c_{t-1} + i_t \otimes \tanh(W_{\tilde{c}x}x_t + W_{\tilde{c}h}h_{t-1})\}\right] \end{aligned} \tag{9.47}$$

$$\begin{aligned} \frac{\partial E}{\partial W_{\tilde{c}h}} &= \frac{\partial E}{\partial y_{p,t}}\frac{\partial y_{p,t}}{\partial h_t}\frac{\partial \mathrm{h_t}}{\partial \mathrm{C^h}}\frac{\partial \mathrm{C^h}}{\partial c_t}\frac{\partial c_t}{\partial \tilde{c}_t}\frac{\partial \tilde{c}_t}{\partial S_c}\frac{\partial S_c}{\partial W_{\tilde{c}h}} \\ &= -\sum_{t=1}^{N}\left(y_{o,t} - y_{p,t}\right)W_{hy}o_t\ \text{dtanh}(c_t)i_t\ \text{dtanh}(S_c)h_{t-1} \end{aligned} \tag{9.48}$$

$$\begin{aligned} \frac{\partial E}{\partial W_{\tilde{c}x}} &= \frac{\partial E}{\partial y_{p,t}}\frac{\partial y_{p,t}}{\partial h_t}\frac{\partial \mathrm{h_t}}{\partial \mathrm{C^h}}\frac{\partial \mathrm{C^h}}{\partial c_t}\frac{\partial c_t}{\partial \tilde{c}_t}\frac{\partial \tilde{c}_t}{\partial S_c}\frac{\partial S_c}{\partial W_{\tilde{c}h}} \\ &= -\sum_{t=1}^{N}\left(y_{o,t} - y_{p,t}\right)W_{hy}o_t\ \text{dtanh}(c_t)i_t\ \text{dtanh}(S_c)x_t \end{aligned} \tag{9.49}$$

where $S_c = W_{\tilde{c}x}x_t + W_{\tilde{c}h}h_{t-1}$.

9.2.4 Backpropagation Through Time (BPTT)

The derivation above does not consider time dependence. The BPTT algorithm must be applied as in RNN. Again, by expanding the simple LSTM in equations from Eqs. (9.26) to (9.31) as

$$y_t = W_{hy}h_t = W_{hy}[o_t \otimes \tanh(c_t)] \tag{9.50}$$

$$y_t = W_{hy}\left[o_t \otimes (f_t \otimes c_{t-1} + i_t \otimes \tilde{c}_t)\right] \tag{9.51}$$

$$\begin{aligned} y_t &= W_{hy}\left[\sigma(W_{ox}x_t + W_{oh}h_{t-1}) \otimes \tanh\left\{\sigma\left(W_{fx}x_t + W_{fh}h_{t-1}\right)\right.\right. \\ &\quad \left.\left. \otimes\, c_{t-1} + \sigma(W_{ix}x_t + W_{ih}h_{t-1}) \otimes \tanh(W_{\tilde{c}x}x_t + W_{\tilde{c}h}h_{t-1})\right]\right. \\ &= W_{hy}\left[\sigma\left(S_{o,t}\right) \otimes \tanh\left\{\sigma\left(S_{f,t}\right) \otimes c_{t-1} + \sigma\left(S_{i,t}\right) \otimes \tanh(S_{\tilde{c},t})\right\}\right] \end{aligned} \tag{9.52}$$

Note that $S_{\delta,t}$ represents the inside term of δ for tanh or sigmoid function. For example, $S_{o,t} = W_{ox}x_t + W_{oh}h_{t-1}$ at $\sigma(W_{ox}x_t + W_{oh}h_{t-1}) = \sigma(S_{o,t})$.

Here, the explicit expression shown in Eq. (9.52) presents 9 weights. Among 9 weights, the derivative of the weight W_{ox} is presented in the following. Further derivation is briefly mentioned at the end of this section.

From these equations, W_{ox} is included in o_t:

$$\begin{aligned} y_t &= W_{hy}h_t = W_{hy}[o_t \otimes \tanh(c_t)] \\ &= W_{hy}\big[[\sigma(\boldsymbol{W_{ox}}x_t + W_{oh}h_{t-1}) \otimes \tanh(c_t)\big] \end{aligned} \tag{9.53}$$

Furthermore, W_{ox} is also included in c_t. In c_t, there are four elements that include W_{ox} as o_t, f_t, i_t, and $\tilde{c}_t$. Each element can explicitly be shown as

$$\begin{aligned} y_t &= W_{hy}h_t = W_{hy}[o_t \otimes \tanh(c_t)] \\ &= W_{hy}\big[o_t \otimes \tanh(f_t \otimes c_{t-1} + i_t \otimes \tilde{c}_t)\big] \end{aligned} \tag{9.54}$$

For o_t,

$$\begin{aligned} y_t &= W_{hy}h_t = W_{hy}[o_t \otimes \tanh(c_t)]] \\ &= W_{hy}\big[\sigma(W_{ox}x_t + W_{oh}h_{t-1}) \otimes \tanh(c_t)\big] \\ &= W_{hy}\big[\sigma\big(W_{ox}x_t + W_{oh}\big\{W_{hy}\big[\sigma(\boldsymbol{W_{ox}}x_{t-1} + W_{oh}h_{t-2}) \otimes \tanh(c_{t-1})\big]\big\}\big) \\ &\quad \otimes \tanh(c_t)\big] \end{aligned} \tag{9.55}$$

For f_t,

$$\begin{aligned} y_t &= W_{hy}h_t = W_{hy}[o_t \otimes \tanh(c_t)] \\ &= W_{hy}\big[o_t \otimes \tanh\big(\sigma\big(W_{fx}x_t + W_{fh}h_{t-1}\big) \otimes c_{t-1} + i_t \otimes \tilde{c}_t\big)\big] \\ &= W_{hy}\big[o_t \otimes \tanh\big(\sigma\big(W_{fx}x_t + W_{fh}\big\{W_{hy}\big[\sigma(\boldsymbol{W_{ox}}x_{t-1} + W_{oh}h_{t-2}) \otimes \tanh(c_{t-1})\big]\big\}\big) \\ &\quad \otimes c_{t-1} + i_t \otimes \tilde{c}_t\big)\big] \end{aligned} \tag{9.56}$$

Also, W_{ox} for i_t,

$$\begin{aligned} y_t &= W_{hy}h_t = W_{hy}[o_t \otimes \tanh(c_t)] = W_{hy}\big[o_t \otimes \tanh(f_t \otimes c_{t-1} + i_t \otimes \tilde{c}_t)\big] \\ &= W_{hy}\big[o_t \otimes \tanh(f_t \otimes c_{t-1} + \sigma(W_{ix}x_t + W_{ih}h_{t-1}) \otimes \tilde{c}_t)\big] \\ &= W_{hy}\big[o_t \otimes \tanh(f_t \otimes c_{t-1} + \sigma(W_{ix}x_t + W_{ih}\big\{W_{hy} \\ &\quad \big[\sigma(\boldsymbol{W_{ox}}x_{t-1} + W_{oh}h_{t-2}) \otimes \tanh(c_{t-1})\big]\big\}\big) \otimes \tilde{c}_t)\big] \end{aligned} \tag{9.57}$$

Also, W_{ox} for $\widetilde{c}_t$,

$$\begin{aligned} y_t &= W_{hy}h_t = W_{hy}[o_t \otimes \tanh(c_t)] = W_{hy}\big[o_t \otimes \tanh(f_t \otimes c_{t-1} + i_t \otimes \tilde{c}_t)\big] \\ &= W_{hy}[o_t \otimes \tanh(f_t \otimes c_{t-1} + i_t \otimes \tilde{c}_t)] \end{aligned}$$

$$= W_{hy}\left[o_t \otimes \tanh\left(f_t \otimes c_{t-1} + i_t \otimes \{\tanh(W_{\tilde{c}x}x_t + W_{\tilde{c}h}h_{t-1})\}\right)\right]$$
$$= W_{hy}\left[o_t \otimes \tanh\left(f_t \otimes c_{t-1} + i_t \otimes \tanh\left(W_{\tilde{c}x}x_t + W_{\tilde{c}h}\left\{W_{hy}\left[\sigma(W_{ox}x_{t-1} + W_{oh}h_{t-2}) \otimes \tanh(c_{t-1})\right]\right\}\right)\right)\right] \quad (9.58)$$

Note that $W_{oh}h_{t-2}$ refers that the derivative must consider not only the current time but also one time ahead.

For example, at time step 1,

$$\frac{\partial E_1}{\partial W_{ox}} = \frac{\partial E_1*}{\partial W_{ox}} + \frac{\partial E_2}{\partial W_{ox}} \quad (9.59)$$

Note that $\frac{\partial E_1*}{\partial W_{ox}}$ is the derivative that only takes the current time into account. These two terms are expressed as below:

$$\frac{\partial E_1*}{\partial W_{ox}} = \frac{\partial E_1}{\partial y_{p,1}} \frac{\partial y_{p,1}}{\partial h_1} \frac{\partial h_1}{\partial o_1(\text{out})} \frac{\partial o_1(\text{out})}{\partial o_1(\text{in})} \frac{\partial o_1(in)}{\partial W_{ox}}$$
$$= \left(y_{o,1} - y_{p,1}\right) W_{hy} \tanh(c_1) \sigma\left(S_{o,1}\right)\left(1 - \sigma\left(S_{o,1}\right)\right) x_1 \quad (9.60)$$

Note that $\frac{\partial o_1(out)}{\partial o_1(in)}$ implies the derivative of o_1 with the inside term represented here with $S_{o,1}$. Subsequently, $\frac{\partial o_1(out)}{\partial o_1(in)}$ becomes $\sigma(S_{o,1})\left(1 - \sigma(S_{o,1})\right)$ as shown in Eq. (4.4). Then,

$$\begin{aligned}\frac{\partial E_2}{\partial W_{oh}} &= \frac{\partial E_2}{\partial y_{p,2}} \frac{\partial y_{p,2}}{\partial h_2} \frac{\partial h_2}{\partial o_2(out)} \frac{\partial o_2(out)}{\partial o_2(in)} \frac{\partial o_2(in)}{\partial h_1} \frac{\partial h_1}{\partial o_1(out)} \frac{\partial o_1(out)}{\partial o_1(in)} \frac{\partial o_1(in)}{\partial W_{ox}} \\ &+ \frac{\partial E_2}{\partial y_{p,2}} \frac{\partial y_{p,2}}{\partial h_2} \frac{\partial h_2}{\partial c_2(out)} \frac{\partial c_2(out)}{\partial c_2(in)} \frac{\partial c_2(in)}{\partial f_2(out)} \frac{\partial f_2(out)}{\partial f_2(in)} \frac{\partial f_2(in)}{\partial h_1} \frac{\partial h_1}{\partial o_1(out)} \frac{\partial o_1(out)}{\partial o_1(in)} \frac{\partial o_1(in)}{\partial W_{ox}} \\ &+ \frac{\partial E_2}{\partial y_{p,2}} \frac{\partial y_{p,2}}{\partial h_2} \frac{\partial h_2}{\partial c_2(out)} \frac{\partial c_2(out)}{\partial c_2(in)} \frac{\partial c_2(in)}{\partial i_2(out)} \frac{\partial i_2(out)}{\partial i_2(in)} \frac{\partial i_2(in)}{\partial h_1} \frac{\partial h_1}{\partial o_1(out)} \frac{\partial o_1(out)}{\partial o_1(in)} \frac{\partial o_1(in)}{\partial W_{ox}} \\ &+ \frac{\partial E_2}{\partial y_{p,2}} \frac{\partial y_{p,2}}{\partial h_2} \frac{\partial h_2}{\partial c_2(out)} \frac{\partial c_2(out)}{\partial c_2(in)} \frac{\partial c_2(in)}{\partial \tilde{c}_2(out)} \frac{\partial \tilde{c}_2(out)}{\partial \tilde{c}_2(in)} \frac{\partial \tilde{c}_2(in)}{\partial h_1} \frac{\partial h_1}{\partial o_1(out)} \frac{\partial o_1(out)}{\partial o_1(in)} \frac{\partial o_1(in)}{\partial W_{ox}}\end{aligned} \quad (9.61)$$

Since this equation is long, each line can be shown separately.

Line $1(o_t)$:

$$\frac{\partial E_2}{\partial y_{p,2}} \frac{\partial y_{p,2}}{\partial h_2} \frac{\partial h_2}{\partial o_2(out)} \frac{\partial o_2(out)}{\partial o_2(in)} \frac{\partial o_2(in)}{\partial h_1} \frac{\partial h_1}{\partial o_1(out)} \frac{\partial o_1(out)}{\partial o_1(in)} \frac{\partial o_1(in)}{\partial W_{ox}}$$
$$= \left(y_{o,2} - y_{p,2}\right) W_{hy} \tanh(c_2) \sigma\left(S_{o,2}\right)\left(1 - \sigma\left(S_{o,2}\right)\right)$$
$$W_{oh} \tanh(c_1) \sigma\left(S_{o,1}\right)\left(1 - \sigma\left(S_{o,1}\right)\right) x_1$$

Line $2(f_t)$:

$$\frac{\partial E_2}{\partial y_{p,2}} \frac{\partial y_{p,2}}{\partial h_2} \frac{\partial h_2}{\partial c_2(out)} \frac{\partial c_2(out)}{\partial c_2(in)} \frac{\partial c_2(in)}{\partial i_2(out)} \frac{\partial i_2(out)}{\partial i_2(in)} \frac{\partial i_2(in)}{\partial h_1} \frac{\partial h_1}{\partial o_1(out)} \frac{\partial o_1(out)}{\partial o_1(in)} \frac{\partial o_1(in)}{\partial W_{ox}}$$
$$= \left(y_{o,2} - y_{p,2}\right) W_{hy} o_2 \left(1 - \tanh(c_2)^2\right) \tilde{c}_2 \sigma\left(S_{i,2}\right)\left(1 - \sigma\left(S_{i,2}\right)\right)$$

$$W_{ih}\tanh(c_1)\sigma\left(S_{o,1}\right)\left(1-\sigma\left(S_{o,1}\right)\right)x_1 \tag{9.63}$$

Line 3(i_t):

$$\begin{aligned}&\frac{\partial E_2}{\partial y_{p,2}}\frac{\partial y_{p,2}}{\partial h_2}\frac{\partial h_2}{\partial c_2(out)}\frac{\partial c_2(out)}{\partial c_2(in)}\frac{\partial c_2(in)}{\partial i_2(out)}\frac{\partial i_2(out)}{\partial i_2(in)}\frac{\partial i_2(in)}{\partial h_1}\frac{\partial h_1}{\partial o_1(out)}\frac{\partial o_1(out)}{\partial o_1(in)}\frac{\partial o_1(in)}{\partial W_{ox}}\\ &\quad=\left(y_{o,2}-y_{p,2}\right)W_{hy}o_2\left(1-\tanh(c_2)^2\right)\tilde{c}_2\sigma\left(S_{i,2}\right)\left(1-\sigma\left(S_{i,2}\right)\right)\\ &\quad W_{ih}\tanh(c_1)\sigma\left(S_{o,1}\right)\left(1-\sigma\left(S_{o,1}\right)\right)x_1\end{aligned} \tag{9.64}$$

Line 4($\widetilde{c}_t$):

$$\begin{aligned}&\frac{\partial E_2}{\partial y_{p,2}}\frac{\partial y_{p,2}}{\partial h_2}\frac{\partial h_2}{\partial c_2(out)}\frac{\partial c_2(out)}{\partial c_2(in)}\frac{\partial c_2(in)}{\partial \tilde{c}_2(out)}\frac{\partial \tilde{c}_2(out)}{\partial \tilde{c}_2(in)}\frac{\partial \tilde{c}_2(in)}{\partial h_1}\frac{\partial h_1}{\partial o_1(out)}\frac{\partial o_1(out)}{\partial o_1(in)}\frac{\partial o_1(in)}{\partial W_{ox}}\\ &\quad=\left(y_{o,2}-y_{p,2}\right)W_{hy}o_2\left(1-\tanh(c_2)^2\right)i_2\left(1-\tanh\left(S_{\tilde{c},2}\right)^2\right)\\ &\quad W_{\tilde{c}h}\tanh(c_1)\sigma\left(S_{o,1}\right)\left(1-\sigma\left(S_{o,1}\right)\right)x_1\end{aligned} \tag{9.65}$$

Overall, the derivative of W_{ox} at time step 1 becomes:

$$\begin{aligned}\frac{\partial E_1}{\partial W_{ox}}&=\frac{\partial E_1}{\partial W_{ox}}+\frac{\partial E_2}{\partial W_{ox}}=\left(y_{o,1}-y_{p,1}\right)W_{hy}\tanh(c_1)\sigma\left(S_{o,1}\right)\left(1-\sigma\left(S_{o,1}\right)\right)x_1\\ &+\ \left(y_{o,2}-y_{p,2}\right)W_{hy}\tanh(c_2)\sigma\left(S_{o,2}\right)\left(1-\sigma\left(S_{o,2}\right)\right)\\ &\quad W_{oh}\tanh(c_1)\sigma\left(S_{o,1}\right)\left(1-\sigma\left(S_{o,1}\right)\right)x_1\\ &+\ \left(y_{o,2}-y_{p,2}\right)W_{hy}o_2\left(1-\tanh(c_2)^2\right)c_1\sigma\left(S_{f,2}\right)\left(1-\sigma\left(S_{f,2}\right)\right)\\ &\quad W_{fh}\tanh(c_1)\sigma\left(S_{o,1}\right)\left(1-\sigma\left(S_{o,1}\right)\right)x_1\\ &+\ \left(y_{o,2}-y_{p,2}\right)W_{hy}o_2\left(1-\tanh(c_2)^2\right)\tilde{c}_2\sigma\left(S_{i,2}\right)\left(1-\sigma\left(S_{i,2}\right)\right)\\ &\quad W_{ih}\tanh(c_1)\sigma\left(S_{o,1}\right)\left(1-\sigma\left(S_{o,1}\right)\right)x_1\\ &+\ \left(y_{o,2}-y_{p,2}\right)W_{hy}o_2\left(1-\tanh(c_2)^2\right)i_2\left(1-\tanh\left(S_{\tilde{c},2}\right)^2\right)\\ &\quad W_{\tilde{c}h}\tanh(c_1)\sigma\left(S_{o,1}\right)\left(1-\sigma\left(S_{o,1}\right)\right)x_1\end{aligned} \tag{9.66}$$

Note that this derivative is only time step 1, while $\frac{\partial E}{\partial W_{ox}}=\sum_{t=1}^{n}\frac{\partial E_t}{\partial W_{ox}}$. For further time steps and other variables,

$$\frac{\partial E_t}{\partial W_.}=\frac{\partial E_t}{\partial y_{p,t}}\frac{\partial y_{p,t}}{\partial h_t}\frac{\partial h_t}{\partial c_t(out)}\frac{\partial c_t(out)}{\partial c_t(in)}\frac{\partial c_t}{\partial c_{t-1}}\frac{\partial c_{t-1}}{\partial c_{t-2}}\frac{\partial c_{t-2}}{\partial c_{t-3}}\cdots\frac{\partial c_1}{\partial c_0}\frac{\partial c_0}{\partial W_.} \tag{9.67}$$

By applying $\frac{\partial c_t}{\partial c_{t-1}}=f_t$

$$\frac{\partial c_t}{\partial c_{t-1}}\frac{\partial c_{t-1}}{\partial c_{t-2}}\frac{\partial c_{t-2}}{\partial c_{t-3}}\cdots\frac{\partial c_1}{\partial c_0}=\prod_{i=1}^{t}f_i \tag{9.68}$$

Note that f_i is the sigmoid function from zero to one and it can minimize the vanishing close to one. High f_i indicates high weight to the previous cell state. It can also avoid exploding, since f_i cannot be bigger than one.

9.3 Gated Recurrent Unit (GRU)

9.3.1 *Basics of GRU*

The Gated Recurrent Unit (GRU) was proposed by Cho et al. (2014) as one of the recurrent neural networks comparable to LSTM. GRU has a simpler architecture than LSTM and employs the gates (update and reset gates) to handle the flow of information in the hidden unit (Fig. 9.3). The performance of GRU has been tested in several studies on comparing LSTM, but no definite conclusion has been made (Chung et al. 2014).

$$\mathbf{r}_t = \sigma(\mathbf{W}_{rx}\mathbf{x}_t + \mathbf{W}_{rh}\mathbf{h}_{t-1})$$

$$\mathbf{z}_t = \sigma(\mathbf{W}_{zx}\mathbf{x}_t + \mathbf{W}_{zh}\mathbf{h}_{t-1})$$

$$\mathbf{h}'_t = tanh(\mathbf{W}_{h'x}\mathbf{x}_t + \mathbf{W}_{h'h}(\mathbf{r}_t \otimes \mathbf{h}_{t-1}))$$

$$\mathbf{h}_t = \mathbf{z}_t\mathbf{h}'_t + (1 - \mathbf{z}_t)\mathbf{h}_{t-1}$$

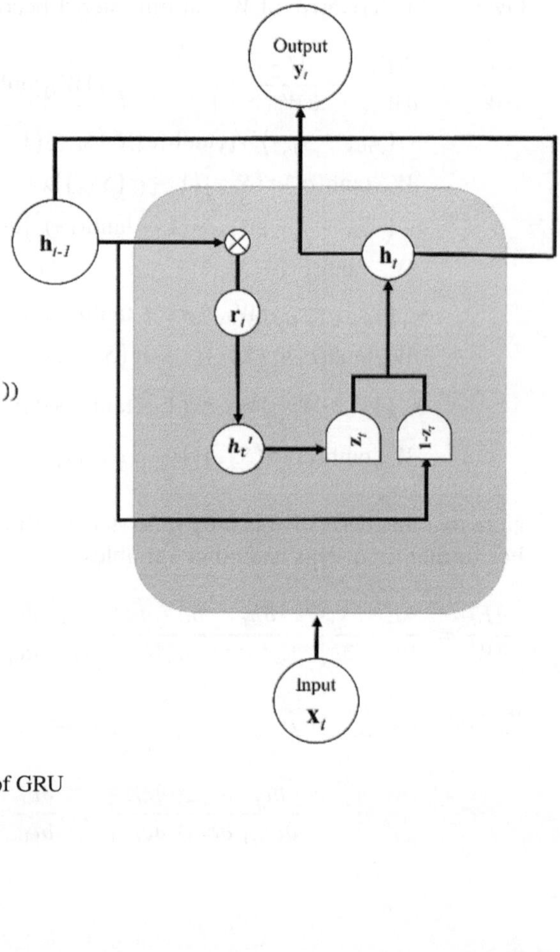

Fig. 9.3 Schematic representation of GRU

The activation of the hidden state of GRU, $\mathbf{h}_t$, at time t is a linear interpolation between the previous hidden state and the candidate hidden state:

$$\mathbf{h}_t = \mathbf{z}_t \otimes \mathbf{h}_{t-1} + (1 - \mathbf{z}_t) \otimes \boldsymbol{h}'_t \tag{9.69}$$

The update gate $\mathbf{z}_t$ and the candidate hidden state are defined as:

$$\mathbf{z}_t = \sigma(\mathbf{W}_{zx}\mathbf{x}_t + \mathbf{W}_{zh}\mathbf{h}_{t-1} + \mathbf{b}_z) \tag{9.70}$$

$$\boldsymbol{h}'_t = \tanh(\mathbf{W}_{h'x}\mathbf{x}_t + \mathbf{W}_{h'h}(\boldsymbol{r}_t \otimes \mathbf{h}_{t-1}) + \mathbf{b}_{h'}) \tag{9.71}$$

where $\boldsymbol{r}_t$ is the reset gate enabling to forget the previous hidden state, denoted as:

$$\boldsymbol{r}_t = \sigma(\mathbf{W}_{rx}\mathbf{x}_t + \mathbf{W}_{rh}\mathbf{h}_{t-1} + \mathbf{b}_r) \tag{9.72}$$

9.3.2 Example of GRU

To present the feedforward GRU procedure, one simple example of GRU is shown here with two hidden units. With the weight parameters in Table 9.4 and the input value and the hidden states shown in Table 9.5, the values of update, reset, and hidden

Table 9.4 Example of GRU weight parameter set with 2 hidden units

		$\mathbf{W}_x$	$\mathbf{W}_h$		b
Hidden 1	$\mathbf{z}_t$	−0.964	−0.099	−0.116	−0.046
	$\mathbf{r}_t$	0.219	0.287	−0.456	0.057
	$\mathbf{h}'_t$	−0.566	0.274	−0.560	−0.020
Hidden 2	$\mathbf{z}_t$	0.086	0.949	1.295	−0.504
	$\mathbf{r}_t$	−0.346	0.791	0.510	−0.351
	$\mathbf{h}'_t$	−0.615	0.289	−0.550	0.023

Table 9.5 Example of the output parameters in the GRU example and dataset

	t−1	t (updated)	NN	
ht	0.5	0.0202	Why1	−0.815
	0.7	0.4203	Why2	−0.550
x_t		0.23	Whyb	−0.084
			Y	−0.3323

Table 9.6 Example of the estimated gates and hidden state updates

		St	g(St)
Hidden 1	$\mathbf{z}_t$	−0.399	0.4016
	$\mathbf{r}_t$	−0.069	0.4828
	$\mathbf{h}'_t$	−0.312	−0.3018
Hidden 2	$\mathbf{z}_t$	0.897	0.7103
	$\mathbf{r}_t$	0.322	0.5798
	$\mathbf{h}'_t$	−0.272	−0.2654

gates were estimated, as shown in Table 9.6. For example, the update gate values $z_t = [z_t^1; z_t^2]$ are estimated as:

$$z_t^1 = \sigma(\boldsymbol{W}_{zx}\boldsymbol{x}_t + \boldsymbol{W}_{zh}\boldsymbol{h}_{t-1} + b_z)$$

$$= \sigma\left(-0.964 \times 0.23 + [-0.099, -0.116]\begin{bmatrix}0.5\\0.7\end{bmatrix} - 0.046\right)$$

$$= \sigma(-0.399) = 1/(1+\exp(-0.399)) = 0.4016$$

$$z_t^2 = \sigma\left(0.086 \times 0.23 + [0.949, 1.295]\begin{bmatrix}0.5\\0.7\end{bmatrix} - 0.504\right)$$

$$= \sigma(0.897) = 0.7103$$

Also, the reset gate ($\mathbf{r}_t$) is estimated as

$$r_t^1 = \sigma(\boldsymbol{W}_{rx}\boldsymbol{x}_t + \boldsymbol{W}_{rh}\boldsymbol{h}_{t-1} + b_r) = 0.4828$$

and $r_t^2 = 0.5798$. These estimates are shown in Table 9.6.

The candidate hidden state $\boldsymbol{h}'_t$ is calculated as:

$$\mathrm{h}_t^{\prime 1} = \tanh(\mathbf{W}_{h'x}\mathbf{x}_t + \mathbf{W}_{h'h}(\boldsymbol{r}_t \otimes \mathbf{h}_{t-1}) + b_{h'}) = \tanh(-0.566 \times 0.23$$

$$+ [0.274, -0.560]\left\{\begin{bmatrix}0.4828\\0.5798\end{bmatrix} \otimes \begin{bmatrix}0.5\\0.7\end{bmatrix}\right\} - 0.02)$$

$$= \tanh(-0.2719) = -0.3018$$

and $\mathrm{h}_t^{\prime 2} = -0.2654$

With these gate values, the hidden state can be updated as:

$$\mathbf{h}_t = \mathbf{z}_t \otimes \mathbf{h}_{t-1} + (1 - \mathbf{z}_t) \otimes \boldsymbol{h}'_t = \begin{bmatrix}0.4016\\0.7103\end{bmatrix} \otimes \begin{bmatrix}0.5\\0.7\end{bmatrix}$$

$$+ \left(1 - \begin{bmatrix}0.4016\\0.7103\end{bmatrix}\right) \otimes \begin{bmatrix}-0.3018\\-0.2654\end{bmatrix} = \begin{bmatrix}0.0202\\0.4203\end{bmatrix}$$

Finally, the predicted value from the GRU model can be estimated with the weight and bias parameter set shown on the right side of Table 9.5 as

$$y_t = \boldsymbol{W}_{hy}\boldsymbol{h}_t + b_{hy} = [-0.815, \ -0.550]\begin{bmatrix} 0.0202 \\ 0.4203 \end{bmatrix} - 0.084 = -0.3323$$

Note that a simple linear activation function is employed as an output activation function.

9.3.3 Backpropagation of a Simple GRU Model

To show the backpropagation algorithm to update the weights of GRU, the system has only one hidden node and only one input without a bias term (i.e., b) as before in the simple LSTM model. Two gates (reset and update) are denoted as:

$$r_t = \sigma(W_{rx}x_t + W_{rh}h_{t-1}) \tag{9.73}$$

$$z_t = \sigma(W_{zx}x_t + W_{zh}h_{t-1}) \tag{9.74}$$

The candidate hidden state and the hidden state are defined as

$$h_t' = \tanh(W_{h'x}x_t + W_{h'h}(r_t \otimes h_{t-1})) \tag{9.75}$$

$$h_t = z_t \otimes h_{t-1} + (1 - z_t) \otimes h_t' \tag{9.76}$$

The output model is simply a linear combination of the hidden state as

$$y_t = W_{hy}h_t \tag{9.77}$$

This model can be explicitly expressed as:

$$\begin{aligned} y_t &= W_{hy}\left(z_t \otimes h_{t-1} + (1 - z_t) \otimes h_t'\right) = W_{hy}\{(1 - (\sigma(W_{zx}x_t + W_{zh}h_{t-1}))) \\ &\quad \otimes h_{t-1} + \sigma(W_{zx}x_t + W_{zh}h_{t-1}) \otimes (\tanh(W_{h'x}x_t + W_{h'h}\sigma(W_{rx}x_t + W_{rh}h_{t-1}) \otimes h_{t-1}))\} \\ &= W_{hy}\left\{\left(1 - \left(\sigma\left(S_{z,t}\right)\right)\right) \otimes h_{t-1} + \sigma\left(S_{z,t}\right) \otimes (\tanh(W_{h'x}x_t + \right. \\ &\quad \left. W_{h'h}\sigma\left(S_{r,t}\right) \otimes h_{t-1}))\right\} \end{aligned} \tag{9.78}$$

This model can be explicitly expressed as:

$$\begin{aligned} y_t &= W_{hy}\left(z_t \otimes h_{t-1} + (1 - z_t) \otimes h_t'\right) \\ &= W_{hy}\{\sigma(W_{zx}x_t + W_{zh}h_{t-1}) \otimes h_{t-1} + (1 - \sigma(W_{zx}x_t + W_{zh}h_{t-1})) \end{aligned}$$

$$\otimes \tanh(W_{h'x}x_t + W_{h'h}\sigma(W_{rx}x_t + W_{rh}h_{t-1}) \otimes h_{t-1})\}$$
$$= W_{hy}\{\sigma(S_{z,t}) \otimes h_{t-1} + (1 - \sigma(S_{z,t})) \otimes \tanh(W_{h'x}x_t$$
$$+ W_{h'h}\sigma(S_{r,t}) \otimes h_{t-1})\} \quad (9.79)$$

In this simple GRU model, 6 parameters are shown. Among 6 parameters, the derivation of the parameter W_{zx} is presented below. The first time derivative has two components as:

$$\frac{\partial E_1}{\partial W_{zx}} = \frac{\partial E_1 *}{\partial W_{zx}} + \frac{\partial E_2}{\partial W_{zx}} \quad (9.80)$$

Note that $\frac{\partial E_1 *}{\partial W_{ox}}$ is the derivative that only takes the current time into account, while $\frac{\partial E_2}{\partial W_{zx}}$ considers the target parameter W_{zx} that is included in the one step forward time.

At first, $\frac{\partial E_1 *}{\partial W_{zx}}$ presents the one that

$$y_t = W_{hy}(z_t \otimes h_{t-1} + (1 - z_t) \otimes h'_t) = W_{hy}(z_t \otimes (h_{t-1} - h'_t) + h'_t)$$
$$= W_{hy}(\sigma(W_{zx}x_t + W_{zh}h_{t-1}) \otimes (h_{t-1} - h'_t) + h'_t) \quad (9.81)$$

$$\frac{\partial E_1}{\partial y_{p,1}} \frac{\partial y_{p,1}}{\partial h_1} \frac{\partial h_1}{\partial z_1(out)} \frac{\partial z_1(out)}{\partial z_1(in)} \frac{\partial z_1(in)}{\partial W_{zx}}$$
$$= (y_{o,1} - y_{p,1})W_{hy}(h_0 - h'_1)\sigma(S_{z,1})(1 - \sigma(S_{z,1}))x_1 \quad (9.82)$$

For the second term, $\frac{\partial E_2}{\partial W_{zx}}$ has four components as follows.

(a) For W_{zx} in z_{t-1} at z_t

$$y_t = W_{hy}(z_t \otimes h_{t-1} + (1 - z_t) \otimes h'_t) = W_{hy}[z_t \otimes (h_{t-1} - h'_t) + h'_t]$$
$$= W_{hy}[(\sigma(W_{zx}x_t + W_{zh}h_{t-1})) \otimes (h_{t-1} - h'_t) + h'_t]$$
$$= W_{hy}[(\sigma(W_{zx}x_t + W_{zh}\{z_{t-1} \otimes (h_{t-2} - h'_{t-1}) + h'_{t-1}\})) \otimes (h_{t-1} - h'_t) + h'_t]$$
$$= W_{hy}[(\sigma(W_{zx}x_t + W_{zh}\{(\sigma(W_{zx}x_{t-1} + W_{zh}h_{t-2})) \otimes (h_{t-2} - h'_{t-1}) + h'_{t-1}\}))$$
$$\otimes(h_{t-1} - h'_t) + h'_t] \quad (9.83)$$

$$\frac{\partial E_2}{\partial y_{p,2}} \frac{\partial y_{p,2}}{\partial h_2} \frac{\partial h_2}{\partial z_2(out)} \frac{\partial z_2(out)}{\partial z_2(in)} \frac{\partial z_2(in)}{\partial h_1} \frac{\partial h_1}{\partial z_1(out)} \frac{\partial z_1(out)}{\partial z_1(in)} \frac{\partial z_1(in)}{\partial W_{zx}}$$
$$= (y_{o,2} - y_{p,2})W_{hy}(h_1 - h'_2)\sigma(S_{z,2})(1 - \sigma(S_{z,2}))W_{zh}(h_0 - h'_1)\sigma(S_{z,1})(1 - \sigma(S_{z,1}))x_1 \quad (9.84)$$

(b) For W_{zx} in h_{t-1} at $z_t \otimes h_{t-1}$,

$$y_t = W_{hy}(z_t \otimes h_{t-1} + (1 - z_t) \otimes h'_t)$$

$$= W_{hy}\big(z_t \otimes \big[z_{t-1} \otimes \big(h_{t-2} - h'_{t-1}\big) + h'_{t-1}\big] + (1 - z_t) \otimes h'_t\big)$$
$$= W_{hy}\big(z_t \otimes \big[\sigma(W_{zx}x_{t-1} + W_{zh}h_{t-2}) \otimes \big(h_{t-2} - h'_{t-1}\big) + h'_{t-1}\big] + (1 - z_t) \otimes h'_t\big) \tag{9.85}$$

$$\frac{\partial E_2}{\partial y_{p,2}}\frac{\partial y_{p,2}}{\partial h_2}\frac{\partial h_2}{\partial h_1}\frac{\partial h_1}{\partial z_1(out)}\frac{\partial z_1(out)}{\partial z_1(in)}\frac{\partial z_1(in)}{\partial W_{zx}}$$
$$= \big(y_{o,2} - y_{p,2}\big)W_{hy}z_2\big(h_0 - h'_1\big)\sigma\big(S_{z,1}\big)\big(1 - \sigma\big(S_{z,1}\big)\big)x_1 \tag{9.86}$$

(c) For W_{zx} in z_{t-1} at h_{t-1} of h'_t

$$y_t = W_{hy}\big(z_t \otimes h_{t-1} + (1 - z_t) \otimes h'_t\big)$$
$$= W_{hy}\big(z_t \otimes h_{t-1} + (1 - z_t) \otimes h'_t\big)$$
$$= W_{hy}\big[z_t \otimes h_{t-1} + (1 - z_t)\tanh(W_{h'x}x_t + W_{h'h}(r_t \otimes h_{t-1}))\big]$$
$$= W_{hy}\big[z_t \otimes h_{t-1} + (1 - z_t)\tanh(W_{h'x}x_t + W_{h'h}(r_t \otimes \{z_{t-1}$$
$$\otimes (h_{t-2} - h'_{t-1}) + h'_{t-1}\})\big]$$
$$= W_{hy}\big[z_t \otimes h_{t-1} + (1 - z_t)\tanh(W_{h'x}x_t + W_{h'h}(r_t \otimes \{\sigma(W_{zx}x_{t-1} + W_{zh}h_{t-2})$$
$$\otimes (h_{t-2} - h'_{t-1} + h'_{t-1}\})\big] \tag{9.87}$$

$$\frac{\partial E_2}{\partial y_{p,2}}\frac{\partial y_{p,2}}{\partial h_2}\frac{\partial h_2}{\partial h'_2(out)}\frac{\partial h'_2(out)}{\partial h'_2(in)}\frac{\partial h'_2(in)}{\partial h_1}\frac{\partial h_1}{\partial z_1(out)}\frac{\partial z_1(out)}{\partial z_1(in)}\frac{\partial z_1(in)}{\partial W_{zx}}$$
$$== \big(y_{o,2} - y_{p,2}\big)W_{hy}(1 - z_2)(1 - \tanh^2\big(S_{h',2}\big))r_2 W_{h'h}\big(h_0 - h'_1\big)$$
$$\sigma\big(S_{z,1}\big)\big(1 - \sigma\big(S_{z,1}\big)\big)x_1 \tag{9.88}$$

(d) For W_{zx} in z_{t-1} at r_t

$$y_t = W_{hy}\big(z_t \otimes h_{t-1} + (1 - z_t) \otimes h'_t\big) = W_{hy}\big(z_t \otimes h_{t-1} + (1 - z_t) \otimes h'_t\big)$$
$$= W_{hy}\big[z_t \otimes h_{t-1} + (1 - z_t)\tanh\big(W_{h'x}x_t + W_{h'h}(r_t \otimes h_{t-1})\big)\big]$$
$$= W_{hy}\big[z_t \otimes h_{t-1} + (1 - z_t)\tanh\big(W_{h'x}x_t + W_{h'h}\big(\sigma(W_{rx}x_t + W_{rh}h_{t-1}) \otimes \ '_{t-1}\big)\big)\big]$$
$$= W_{hy}\big[z_t \otimes h_{t-1} + (1 - z_t)\tanh\big(W_{h'x}x_t + W_{h'h}(\sigma(W_{rx}x_t$$
$$+ W_{rh}\big\{\sigma\big(W_{zx}x_{t-1} + W_{zh}h_{t-2}\big) \otimes \big(h_{t-2} - h'_{t-1}\big) + h'_{t-1}\big\} \otimes \big(h_{t-1}\big)\big)\big] \tag{9.89}$$

$$\frac{\partial E_2}{\partial y_{p,2}}\frac{\partial y_{p,2}}{\partial h_2}\frac{\partial h_2}{\partial h'_2(out)}\frac{\partial h'_2(out)}{\partial h'_2(in)}\frac{\partial h'_2(in)}{\partial r_2(out)}\frac{\partial r_2(out)}{\partial r_2(in)}\frac{\partial r_2(in)}{\partial h_1}\frac{\partial h_1}{\partial z_1(out)}\frac{\partial z_1(out)}{\partial z_1(in)}\frac{\partial z_1(in)}{\partial W_{zx}}$$
$$== \big(y_{o,2} - y_{p,2}\big)W_{hy}(1 - z_2)(1 - \tanh^2\big(S_{h',2}\big))h_1 W_{h'h}\big(h_0 - h'_1\big)\sigma\big(S_{z,1}\big)\big(1 - \sigma\big(S_{z,1}\big)\big)x_1 \tag{9.90}$$

Finally,

$$\frac{\partial E_1}{\partial W_{zx}} = \frac{\partial E_1 *}{\partial W_{zx}} + \frac{\partial E_2}{\partial W_{zx}} = \frac{\partial E_1}{\partial y_{p,1}}\frac{\partial y_{p,1}}{\partial h_1}\frac{\partial h_1}{\partial z_1(out)}\frac{\partial z_1(out)}{\partial z_1(in)}\frac{\partial z_1(in)}{\partial W_{zx}}$$
$$+ \frac{\partial E_2}{\partial y_{p,2}}\frac{\partial y_{p,2}}{\partial h_2}\frac{\partial h_2}{\partial z_2(out)}\frac{\partial z_2(out)}{\partial z_2(in)}\frac{\partial z_2(in)}{\partial h_1}\frac{\partial h_1}{\partial z_1(out)}\frac{\partial z_1(out)}{\partial z_1(in)}\frac{\partial z_1(in)}{\partial W_{zx}}$$
$$+ \frac{\partial E_2}{\partial y_{p,2}}\frac{\partial y_{p,2}}{\partial h_2}\frac{\partial h_2}{\partial h_1}\frac{\partial h_1}{\partial z_1(out)}\frac{\partial z_1(out)}{\partial z_1(in)}\frac{\partial z_1(in)}{\partial W_{zx}}$$

$$+\frac{\partial E_2}{\partial y_{p,2}}\frac{\partial y_{p,2}}{\partial h_2}\frac{\partial h_2}{\partial h_2'(out)}\frac{\partial h_2'(out)}{\partial h_2'(in)}\frac{\partial h_2'(in)}{\partial h_1}\frac{\partial h_1}{\partial z_1(out)}\frac{\partial z_1(out)}{\partial z_1(in)}\frac{\partial z_1(in)}{\partial W_{zx}}$$
$$+\frac{\partial E_2}{\partial y_{p,2}}\frac{\partial y_{p,2}}{\partial h_2}\frac{\partial h_2}{\partial h_2'(out)}\frac{\partial h_2'(out)}{\partial h_2'(in)}\frac{\partial h_2'(in)}{\partial r_2(out)}\frac{\partial r_2(out)}{\partial r_2(in)}\frac{\partial r_2(in)}{\partial h_1}\frac{\partial h_1}{\partial z_1(out)}\frac{\partial z_1(out)}{\partial z_1(in)}\frac{\partial z_1(in)}{\partial W_{zx}} \tag{9.91}$$

$$\begin{aligned}\frac{\partial E_1}{\partial W_{zx}} &= \frac{\partial E_1*}{\partial W_{zx}}+\frac{\partial E_2}{\partial W_{zx}} = \frac{\partial E_1}{\partial y_{p,1}}\frac{\partial y_{p,1}}{\partial h_1}\frac{\partial h_1}{\partial z_1(out)}\frac{\partial z_1(out)}{\partial z_1(in)}\frac{\partial z_1(in)}{\partial W_{zx}}\\
&+\frac{\partial E_2}{\partial y_{p,2}}\frac{\partial y_{p,2}}{\partial h_2}\frac{\partial h_2}{\partial z_2(out)}\frac{\partial z_2(out)}{\partial z_2(in)}\frac{\partial z_2(in)}{\partial h_1}\frac{\partial h_1}{\partial z_1(out)}\frac{\partial z_1(out)}{\partial z_1(in)}\frac{\partial z_1(in)}{\partial W_{zx}}\\
&+\frac{\partial E_2}{\partial y_{p,2}}\frac{\partial y_{p,2}}{\partial h_2}\frac{\partial h_2}{\partial h_1}\frac{\partial h_1}{\partial z_1(out)}\frac{\partial z_1(out)}{\partial z_1(in)}\frac{\partial z_1(in)}{\partial W_{zx}}\\
&+\frac{\partial E_2}{\partial y_{p,2}}\frac{\partial y_{p,2}}{\partial h_2}\frac{\partial h_2}{\partial h'_2(out)}\frac{\partial h'_2(out)}{\partial h'_2(in)}\frac{\partial h'_2(in)}{\partial h_1}\frac{\partial h_1}{\partial z_1(out)}\frac{\partial z_1(out)}{\partial z_1(in)}\frac{\partial z_1(in)}{\partial W_{zx}}\\
&+\frac{\partial E_2}{\partial y_{p,2}}\frac{\partial y_{p,2}}{\partial h_2}\frac{\partial h_2}{\partial h'_2(out)}\frac{\partial h'_2(out)}{\partial h'_2(in)}\frac{\partial h'_2(in)}{\partial r_2(out)}\frac{\partial r_2(out)}{\partial r_2(in)}\frac{\partial r_2(in)}{\partial h_1}\\
&\frac{\partial h_1}{\partial z_1(out)}\frac{\partial z_1(out)}{\partial z_1(in)}\frac{\partial z_1(in)}{\partial W_{zx}}\end{aligned}$$

Then,

$$\begin{aligned}\frac{\partial E_1}{\partial W_{zx}} &= (y_{o,1}-y_{p,1})W_{hy}(h_0-h_1')\sigma(S_{z,1})(1-\sigma(S_{z,1}))x_1\\
&+ (y_{o,2}-y_{p,2})W_{hy}(h_1-h_2')\sigma(S_{z,2})(1-\sigma(S_{z,2}))\\
&W_{zh}(h_0-h_1')\sigma(S_{z,1})(1-\sigma(S_{z,1}))x_1\\
&+ (y_{o,2}-y_{p,2})W_{hy}z_2(h_0-h_1')\sigma(S_{z,1})(1-\sigma(S_{z,1}))x_1+(y_{o,2}-y_{p,2})\\
&W_{hy}(1-z_2)\left(1-\tanh^2(S_{h',2})\right)r_2W_{h'h}(h_0-h_1')\sigma(S_{z,1})(1-\sigma(S_{z,1}))x_1\\
&+ (y_{o,2}-y_{p,2})W_{hy}(1-z_2)\left(1-\tanh^2(S_{h',2})\right)h_1W_{h'h}(h_0-h_1')\sigma(S_{z,1})(1-\sigma(S_{z,1}))x_1\end{aligned} \tag{9.92}$$

For further time steps, since $\frac{\partial E}{\partial W_{zx}} = \sum_{t=1}^{n}\frac{\partial E_t}{\partial W_{zx}}$,

$$\frac{\partial E_t}{\partial W_{zx}} = \frac{\partial E_t}{\partial y_t}\frac{\partial y_t}{\partial h_t}\frac{\partial h_t}{\partial W_{zx}} \tag{9.93}$$

$$\frac{\partial h_t}{\partial W_{zx}} = \sum_{j=1}^{t}\frac{\partial h_t}{\partial h_i}\frac{\partial h_i}{\partial W_{zx}} = \sum_{j=1}^{t}\left(\prod_{i=j}^{t-1}\frac{\partial h_{i+1}}{\partial h_i}\right)\frac{\partial h_i}{\partial W_{zx}} \tag{9.94}$$

Here, note that

$$\begin{aligned}h_t = z_t\otimes h_{t-1}+(1-z_t)\otimes h_t' &= \sigma(W_{zx}x_t+W_{zh}h_{t-1})\\
&\otimes\ h_{t-1}+(1-\sigma(W_{zx}x_t+W_{zh}h_{t-1}))\end{aligned}$$

$$\otimes \tanh(W_{h'x}x_t + W_{h'h}\sigma(W_{rx}x_t + W_{rb}h_{t-1}) \otimes h_{t-1}) \tag{9.95}$$

Therefore,

$$\begin{aligned}
\frac{\partial h_t}{\partial h_{t-1}} &= \frac{\partial h_t}{\partial h't}\frac{\partial h'_t}{\partial h_{t-1}} + \frac{\partial h_t}{\partial z_t}\frac{\partial z_t}{\partial h_{t-1}} + \frac{\partial h^*_t}{\partial h_{t-1}} \\
&= \frac{\partial h_t}{\partial h't}\left(\frac{\partial h'_t}{\partial r_t}\frac{\partial r_t}{\partial h_{t-1}} + \frac{\partial h_t}{\partial h_{t-1}}\right) + \frac{\partial h_t}{\partial z_t}\frac{\partial z_t}{\partial h_{t-1}} + \frac{\partial h^*_t}{\partial h_{t-1}} \\
&= (1 - z_t)\Big\{(1 - \tanh^2(S_{h',t})\Big)W_{h'h}h_{t-1}(1 - r_t)r_t W_{rh} + \Big(1 - \tanh^2(S_{h',t})\Big)W_{h'h}r_t\Big\} \\
&\quad W_{h'h}r_t + (h_{t-1} - h'_t)z_t(1 - z_t)W_{zh} + z_t
\end{aligned} \tag{9.96}$$

Note that $\frac{\partial h^*_t}{\partial h_{t-1}}$ is the gradient of h_t with letting h'_t and z_t be constant.

9.4 Exercises

1. Fit the RNN model to the Colorado River data in Table 2.3 but with no bias (i.e., only u_1 and w_1).
2. Derive $\frac{\partial E_1}{\partial W_{ix}}$ with BPTT for the LSTM model.
3. Estimate the output of the LSTM model for the Colorado River data in Table 2.3. Here, employ the parameter set of Table 9.1. Set the first value of hidden and cell states are zero.
4. Derive $\frac{\partial E_1}{\partial W_{zh}}$ with backpropagation and BPTT for the GRU model.
5. Estimate the output of the GRU model for the Colorado River data in Table 2.3. Here, employing the parameter set of Table 9.4. Set the first value of hidden and cell states as zero.

References

Cho K et al (2014) Learning phrase representations using RNN encoder-decoder for statistical machine translation. https://arxiv.org/pdf/1406.1078v3.pdf

Chung J, Gulcehre C, Cho K, Bengio Y (2014) Empirical evaluation of gated recurrent neural networks on sequence modeling. https://arxiv.org/abs/1412.3555

Hochreiter S, Schmidhuber J (1997) Long short-term memory. Neural Comput 9(8):1735–1780

Chapter 10
Deep Learning for Spatial Datasets

Abstract Among recent developments of deep learning models, the availability of spatial datasets or images with deep learning is the most significant contribution. In this chapter, a convolutional neural network (CNN) that can analyze the spatial datasets is described. The training procedure of CNN is described with a simple example.

10.1 Convolutional Neural Network (CNN)

10.1.1 Definition of Convolution

Convolution

Convolution is a mathematical operation that integrates the product of two functions or filters (e.g., f and g) with one of them flipped, denoted as

$$(f * g)(t) = \int_{-\infty}^{\infty} f(\tau)g(t-\tau)d\tau \tag{10.1}$$

where * is the convolution operator.

In a discrete case,

$$(f * g)(m) = \sum_{n} f(n)g(m-n) \tag{10.2}$$

In a hydrologic system, rainfall and runoff system can be described with a convolution. With a unit effective rainfall amount (1 cm or 1 inch), the direct runoff hydrograph is defined as a unit hydrograph (UH, $h(t)$). As shown in Fig. 10.6, the runoff from a rainfall distribution denoted as $I(\tau)$ can be described with convolution as

T. Lee et al., *Deep Learning for Hydrometeorology and Environmental Science*, Water Science and Technology Library 99,
https://doi.org/10.1007/978-3-030-64777-3_10

$$Q(t) = \int_{-\infty}^{\infty} I(\tau)h(t - \tau)d\tau \tag{10.3}$$

The runoff discharge at time t, $(Q(t))$, can be estimated by integrating the product of the rainfall amount at time τ and the flipped unit hydrograph at time $t - \tau$. The unit hydrograph is flipped since the unit runoff discharge that contributes to the total runoff discharge at t $(Q(t))$ occurs from the rainfall at time τ, $I(\tau)$ arriving at time $t - \tau$ (Fig. 10.1).

As an example of convolution in a neural network, consider a 5 × 5 input feature map $(I(j, k), j, k = 1,\ldots,5)$ and a 3 × 3 filter $(F(a, b), a, b = 1,\ldots,3)$. The convolution output is denoted as C(l, m). The convolution output is estimated as follows. For $\boldsymbol{C}(\mathbf{1}, \mathbf{1})$,

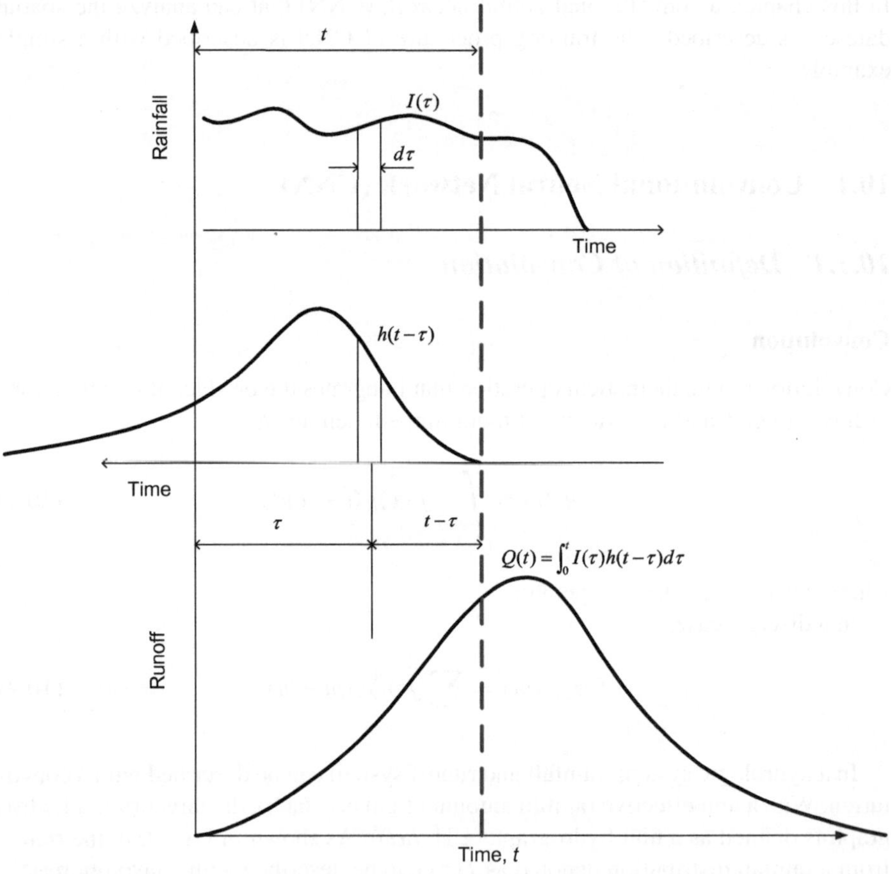

Fig. 10.1 Representation of convolution concept: rainfall (top panel), IUH (middle panel), runoff (bottom panel)

$$\boldsymbol{C(1,1) = F_{1,1}I_{1,1} + F_{1,2}I_{1,2} + F_{1,3}I_{1,3} + \cdots + F_{3,3}I_{3,3}}$$

The schematic diagram is shown in the top part of Fig. 10.2. The two-dimensional filter is multiplied over the input feature map starting from the top left corner and moves one step right and goes up to the end and the next row, etc. The following one is expressed as:

$$\begin{bmatrix} \boldsymbol{C(1,1) = F_{1,1}I_{1,1} + F_{1,2}I_{1,2} + F_{1,3}I_{1,3} + \cdots + F_{3,3}I_{3,3}} \\ C(1,2) = F_{1,1}I_{1,2} + F_{1,2}I_{1,3} + F_{1,3}I_{1,4} + \cdots + F_{3,3}I_{3,4} \\ C(1,3) = F_{1,1}I_{1,3} + F_{1,2}I_{1,4} + F_{1,3}I_{1,5} + \cdots + F_{3,3}I_{3,5} \\ C(2,1) = F_{1,1}I_{2,1} + F_{1,2}I_{2,2} + F_{1,3}I_{2,3} + \cdots + F_{3,3}I_{4,3} \\ \vdots \\ C(3,3) = F_{1,1}I_{3,3} + F_{1,2}I_{3,4} + F_{1,3}I_{3,5} + \cdots + F_{3,3}I_{5,5} \end{bmatrix}$$

This convolution layer ($C(l,m)$) for the ith data can be simply described as

$$C_i(l,m) = \sum_{a=1}^{N_{FH}} \sum_{b=1}^{N_{FW}} F_{a,b}(i) I_{l+a-1,m+b-1}(i) \tag{10.4}$$

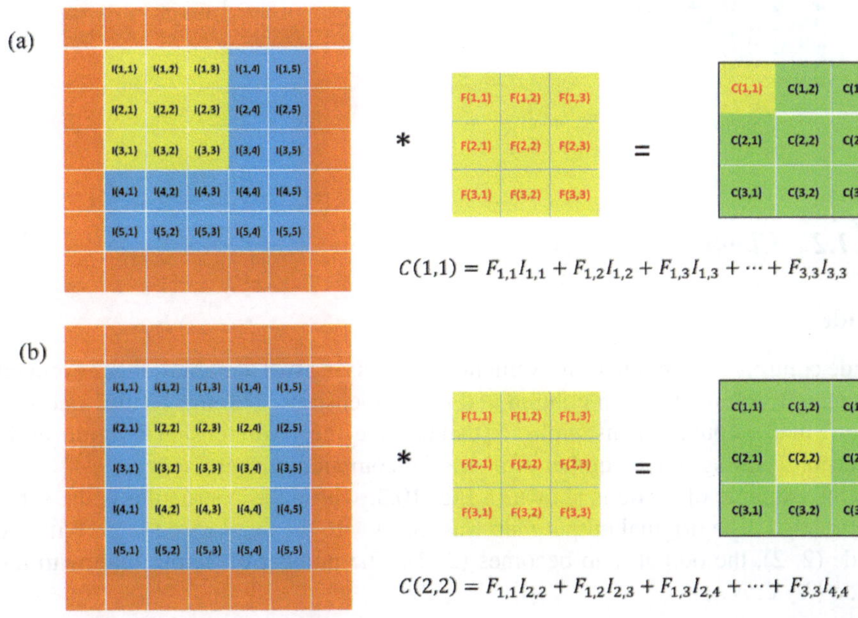

Fig. 10.2 Example of convolving an input feature map (I) and a filter (F) into an output map (C) for C(1,1) in panel **(a)** and for C(2, 2) in panel **(b)**

where *FH* and *FW* are the number of the filter height and width, respectively (here, $N_{FH} = 3$ and $N_{FW} = 3$). The final resolution of the convolution layer (N_{CW}, N_{CH}) becomes

$$N_{CW} = N_{IW} - N_{FW} + 1 \quad \text{and} \quad N_{CH} = N_{IH} - N_{FH} + 1 \tag{10.5}$$

For example, here N_{IH} and $N_{IH} = 5$, and N_{FW} and $N_{FH} = 3$. Therefore, $(N_{CW}, N_{CH}) = (5–3 + 1,5–3 + 1) = (3,3)$.

Example 10.1 Estimate the convolution layer for the input and the filter below.

Solution:

$$C(1,1) = \mathbf{6 \times 0 + 6 \times 0 + 6 \times 0 + 5 \times 1 \cdots + 3 \times 0 = 17.0}$$

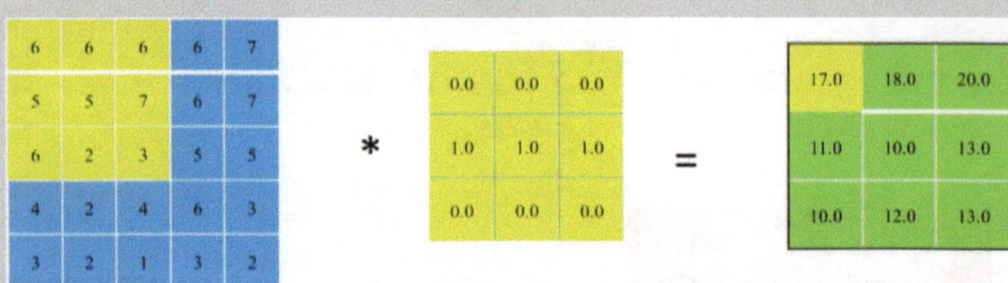

10.1.2 Elements of CNN

Stride

Stride controls the amount of movement in convolving two variables; the movement is generally symmetrical in column and row dimensions. Through stride, a normal feature map output is downsampled so that the size of each dimension is significantly reduced, resulting in the number of pixels for computational efficiency.

One example of stride is shown in Fig. 10.3, whose basic structure is the same as Fig. 10.2. The original output map was (3, 3) with the input map (5, 5). With the stride (2, 2), the output map becomes (2, 2). The number of output map with the stride (N_{SW}, N_{SH})

$$N_{CW} = \langle (N_{IW} - N_{FW} + N_{SW} + 1)/N_{SW} \rangle \tag{10.6}$$

and

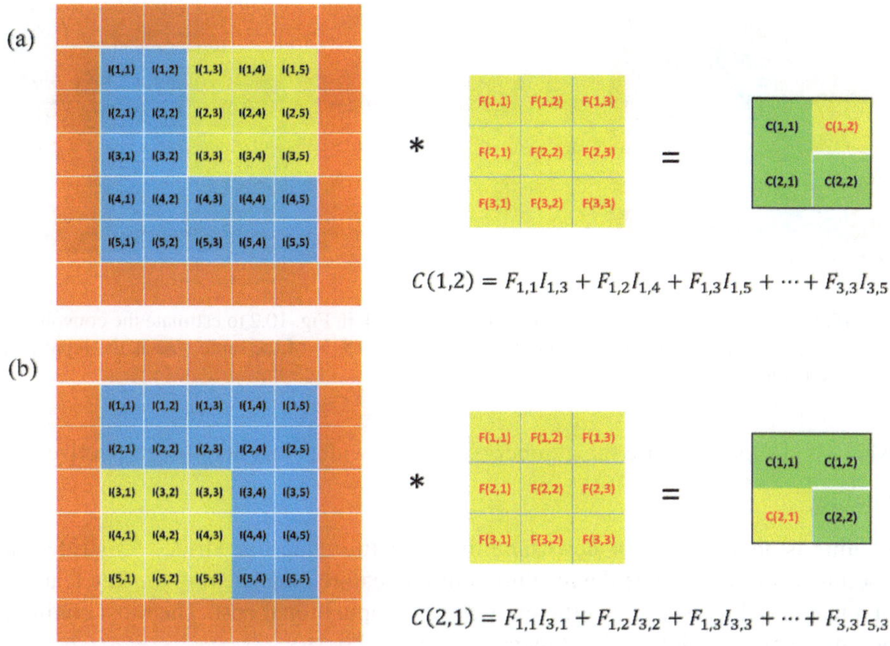

Fig. 10.3 Example of "stride" of (2, 2) in a convolution network in Fig. 10.2 to estimate the convolution output map of $C(1, 2)$ and $C(2, 1)$

$$N_{CH} = \langle (N_{IH} - N_{FH} + N_{SH} + 1)/N_{SH} \rangle \tag{10.7}$$

where < . > indicates the floor (greatest preceding integer).

For example, in Fig. 10.3 with the stride (2, 2):

$$N_{CW} = \langle (N_{IW} - N_{FW} + N_{SW} + 1)/N_{SW} \rangle = \langle (5 - 3 + 2 + 1)/2 \rangle = \langle 3.5 \rangle = 3$$

Padding

Padding is a term describing the process of filling a field with pad characters, here zero. Without padding, the size of the input layer decreases in the output layer. For example, as in Fig. 10.2, the input size (5, 5) becomes (3, 3) in the output layer. Padding the surrounding area with zero values avoids the decrease of the size, as shown in the example of Fig. 10.4. After padding, the output size changes to (5, 5), the same size as the input layer. The size of a convolution output layer can be calculated as

$$N_{CW} = N_{IW} - N_{FW} + N_{PW} + 1 \text{ and } N_{CH} = N_{IH} - N_{FH} + N_{PH} + 1 \tag{10.8}$$

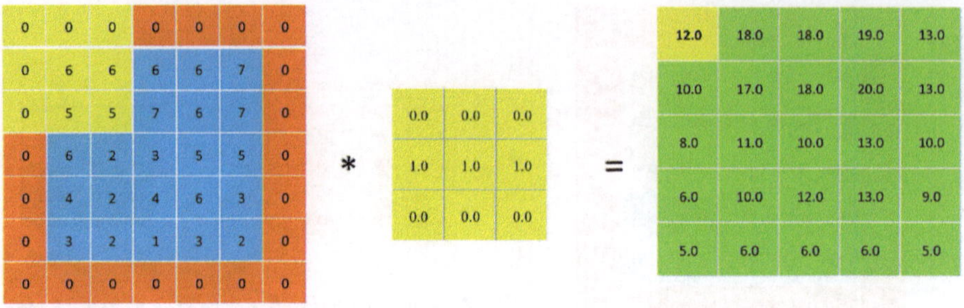

Fig. 10.4 Example of "padding" in a convolution network in Fig. 10.2 to estimate the convolution output map of $C(1,1)$. Note that the output map becomes (5, 5) of the same size as the input map on the left side

where N_{PW} and N_{PH} are the number of pooling widths and heights, respectively.

Pooling

Pooling is another way of downsampling a feature map like stride. As stride, the pooling layer is to reduce the resolution of the feature map keeping the key features of a map. Pooling is also an efficient way of computational cost. The most common pooing techniques are max and mean pooling.

In Fig. 10.5, the example of max and mean pooling is shown. The (4, 4) size input map results in the (2, 2) output map through (2, 2) max pooling. Max pooling obtains the largest value among the pooling window (here, 2 × 2), while mean pooling takes on the average over the values in the pooling window. The size of the output map

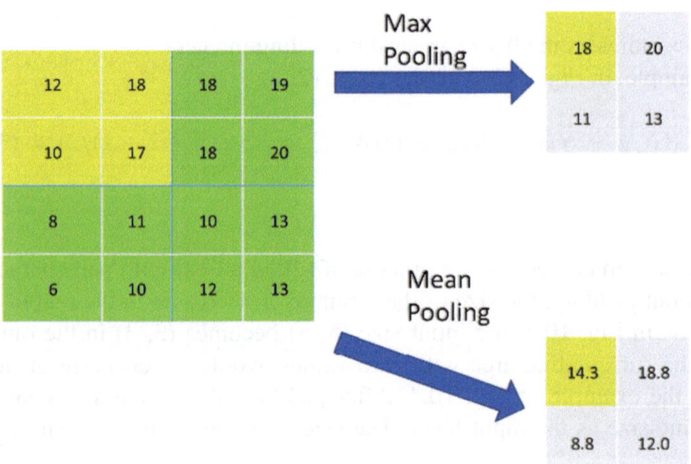

Fig. 10.5 Example of max (right-top) and mean (right bottom) pooling in the convolution network in Fig. 10.2 to estimate the convolution output map of $C(1, 1)$. Note that the output map becomes (5, 5) of the same size as the input map on the left side

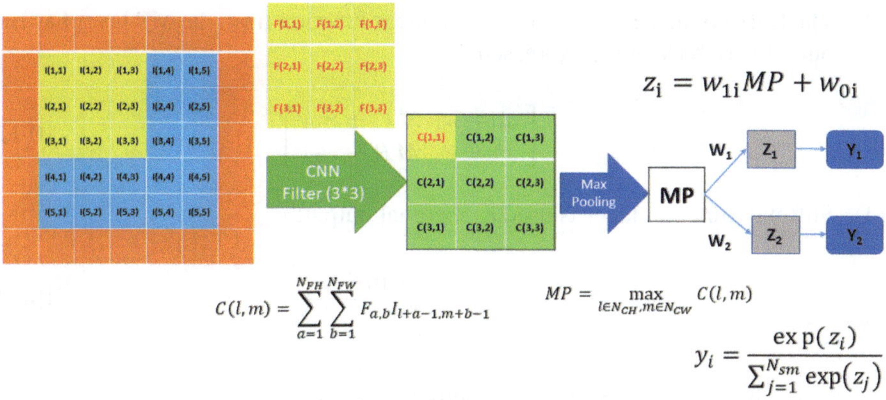

Fig. 10.6 Example of a simple Convolution Neural Network (CNN)

for pooling is $\langle (N_{IW})/N_{PW} \rangle$, where N_{PW} is the number of the pooling width. For example, the output map by pooling in Fig. 10.5 is

$$\langle (N_{IW})/N_{PW} \rangle = \langle 4/2 \rangle = \langle 2 \rangle = 2$$

10.2 Backpropagation of CNN

CNN contains a huge number of parameters to estimate, especially for filters ($F(a, b)$). Following the same way in a regular neural network, the backpropagation with the gradient descent method is commonly employed to estimate the parameters.

In the following, one simple example of CNN is provided, as shown in Fig. 10.6. In the current example, the input map data with (5, 5) size is used, and just one (3, 3) filter ($F(a, b)$) convolves the convolution layer with (3, 3). The convolution map is pooled with Max Pooling (3, 3), resulting in one output variable. The output variable is transferred into the probability distribution with the Softmax function Eq. (4.19) for classification. The employed four layers are shown as:

(1) Convolution layer ($C(l,m)$) is:

$$C_i(l, m) = \sum_{a=1}^{N_{FH}} \sum_{b=1}^{N_{FW}} F_{a,b}(i) I_{l+a-1,m+b-1}(i) \tag{10.9}$$

(2) Max Pooling layer is applied, resulting in one variable as

$$MP_i = \max_{l \in N_{CH}, m \in N_{CW}} C_i(l, m) \tag{10.10}$$

(3) The fully connected layer (here with linear function from the MP layer has two output variables and is expressed as

$$\begin{bmatrix} z_i^1 = w_{11}MP_i + w_{01} \\ z_i^2 = w_{12}MP + w_{02} \end{bmatrix} \tag{10.11}$$

(4) Softmax function layer is used as the final output:

$$y_{p,i}^c = \frac{\exp(z_i^c)}{\sum_{c'=1}^{N_{sm}} \exp(z_i^{c'})} \tag{10.12}$$

where N_{sm} is the number of Softmax outputs, here $N_{sm} = 2$.

For the loss function with the cross-entropy-error as:

$$L = \sum_{i=1}^{N} E_i = -\sum_{i=1}^{N}\sum_{c=1}^{N_{sm}} y_{o,i}^c \log(y_{p,i}^c) \tag{10.13}$$

Finally, the output value of CNN can be described as:

$$y_{p,i}^c = \frac{\exp(z_i^c)}{S} = \frac{\exp(w_{1c}MP_i + w_{0c})}{S} = \frac{\exp\left[w_{1c}\left\{\max_{l\in N_{CH}, m\in N_{CW}} C_i(l,m)\right\} + w_{0c}\right]}{S}$$
$$= \frac{\exp\left[w_{1c}\left\{\max_{l\in N_{CH}, m\in N_{CW}} \sum_{a=1}^{N_{FH}}\sum_{b=1}^{N_{FW}}\left(F_{a,b}(i) I_{l+a-1,m+b-1}(i)\right)\right\} + w_{0c}\right]}{S} \tag{10.14}$$

where S $= \sum_{c'=1}^{N_{sm}} exp(z_i^{c'})$.

In this CNN example, the employed parameters are w_{1i} and w_{0i} as well as all the filters. To update the parameters with the backpropagation algorithm, their derivatives must be estimated.

For the output variable with CEE,

$$\frac{\partial E_i}{\partial y_{p,i}^c} = -\frac{y_{o,i}^c}{y_{p,i}^c} \tag{10.15}$$

Subsequently, the derivative of the Softmax function layer z_i^c is denoted as:

$$\frac{\partial y_{p,i}^c}{\partial z_i^c} = \frac{\exp(z_i^c)S - \exp(z_i^c)^2}{S^2} = y_{p,i}^c(1 - y_{p,i}^c) \tag{10.16}$$

where

$$\frac{\partial\left(\frac{f(x)}{g(x)}\right)}{\partial x}=\frac{f\prime(x)g(x)-f(x)g\prime(x)}{g(x)^2} \tag{10.17}$$

Note that the case of $c \neq c\prime$ must be considered, including the case of $c = c\prime$ in Eq. (10.16) as:

$$\frac{\partial y^{c\prime}_{p,i}}{\partial z^c_i}=\frac{0-\exp\left(z^c_i\right)\exp\left(z^{c\prime}_i\right)}{S^2}=\frac{0-\exp\left(z^c_i\right)\exp\left(z^{c\prime}_i\right)}{S^2}=-y^c_{p,i}y^{c\prime}_{p,i} \tag{10.18}$$

Here, for all $\frac{\partial L}{\partial y^c_{p,i}}\frac{\partial y^{c\prime}_{p,i}}{\partial z^c_i}$, that is, $c \neq c\prime$ and $c = c\prime$

$$\begin{aligned}\frac{\partial L}{\partial y^c_{p,i}}\frac{\partial y^c_{p,i}}{\partial z^c_i}&=-\frac{y^c_{o,i}}{y^c_{p,i}}\left(y^c_{p,i}\left(1-y^c_{p,i}\right)\right)+\frac{y^{c'}_{o,i}}{y^{c'}_{p,i}}y^c_{p,i}y^{c'}_{p,i}=-y^c_{o,i}\left(1-y^c_{p,i}\right)-y^{c'}_{o,i}y^c_{p,i}\\&=-y^c_{o,i}+y^c_{o,i}y^c_{p,i}-y^{c'}_{o,i}y^c_{p,i}=-y^c_{o,i}+y^c_{p,i}\left(y^c_{o,i}+y^{c'}_{o,i}\right)=y^c_{p,i}-y^c_{o,i}\end{aligned} \tag{10.19}$$

where, always $y^c_{o,i}+y^{c'}_{o,i}=1$. Note that the derivative result for the combined CEE and Softmax function can be easily described as $\frac{\partial L}{\partial y^c_{p,i}}\frac{\partial y^c_{p,i}}{\partial z^c_i}=y^c_{p,i}-y^c_{o,i}$. This expression is also an additional reason that CEE is frequently employed in the classification with the Softmax function.

For the weights of the fully connected layer with $z^c_i = w_{1c}MP_i + w_{0c}$,

$$\frac{\partial z^c_i}{\partial w_{1c}}=MP_i \tag{10.20}$$

$$\frac{\partial z^c_i}{\partial w_{0c}}=1 \tag{10.21}$$

Therefore, the derivatives (or gradients) of the weights are denoted as

$$\frac{\partial L}{\partial w_{1c}}=\frac{\partial L}{\partial y^c_{p,i}}\frac{\partial y^c_{p,i}}{\partial z^c_i}\frac{\partial z^c_i}{\partial w_{1c}}=\left(y^c_{p,i}-y^c_{o,i}\right)MP_i \tag{10.22}$$

$$\frac{\partial L}{\partial w_{0c}}=\frac{\partial L}{\partial y^c_{p,i}}\frac{\partial y^c_{p,i}}{\partial z^c_i}\frac{\partial z^c_i}{\partial w_{0c}}=\left(y^c_{p,i}-y^c_{o,i}\right) \tag{10.23}$$

For the filter parameters, the derivative of $\frac{\partial z^c_i}{\partial MP}$ and $\frac{\partial MP_i}{\partial C_i}$ must be derived as:

$$\frac{\partial z_i^c}{\partial MP} = w_{1c} \tag{10.24}$$

$$\frac{\partial MP_i}{\partial C_i} = \begin{cases} 1 \ for \ \max C_i(l, m) \\ 0 \ otherwise \end{cases} \tag{10.25}$$

The derivative of the filter parameters is derived as

$$\frac{\partial C_i}{\partial F_{a,b}} = I_{l+a-1,m+b-1} \tag{10.26}$$

Finally, the full derivative of the filter parameters to CEE is:

$$\begin{aligned} \frac{\partial E_i}{\partial F_{a,b}} &= \sum_{c=1}^{2} \frac{\partial L}{\partial y_{p,i}^c} \frac{\partial y_{p,i}^c}{\partial z_i^c} \frac{\partial z_i^c}{\partial MP_i} \frac{\partial MP_i}{\partial C_i} \frac{\partial C_i}{\partial F_{a,b}} \\ &= \sum_{c=1}^{2} \left(y_{p,i}^c - y_{o,i}^c\right) w_{1c} \left(I_{l+a-1,m+b-1} \ for \ \max C_i(l, m)\right) \end{aligned} \tag{10.27}$$

The derivative of the loss function is produced with the summation of the derivative of CEE for $i = 1,\ldots, n$, that is,

$$\frac{\partial L}{\partial w_{j,k}} = \sum_{i=1}^{n} \frac{\partial E_i}{\partial w_{j,k}} \tag{10.28}$$

$$\frac{\partial L}{\partial F_{a,b}} = \sum_{i=1}^{n} \frac{\partial E_i}{\partial F_{a,b}} \tag{10.29}$$

Example 10.2 Consider the same network structure as in Fig. 10.6 and the spatial dataset of mean sea level pressure (MSLP) with (5, 5) input layer (see Table 10.1) and the final output data defining whether it is northerly wind or not, according to the MLSP.

In this example, defining northerly wind or not (1 or 0, respectively) can be done by comparing the upper and lower part of MLSP. The upper row values of MSLP data in Table 10.1 must be higher than the lower row values to have northerly wind. The data in Table 10.1 clearly shows that the upper row values of MSLP is higher than the lower row values indicating the northerly wind. The indication of the northerly wind must be made in advance (e.g., from weather forecaster) and this indication dataset is used as the final output (i.e., y_o). Since the current example is for classification, one-hot encoding must be applied as [1, 0] indicating that the current data in Table 10.1 is northerly wind and one

Table 10.1 Example MSLP dataset for the example of CNN in Example 10.2

Data	1	2	3	4	5
1	999.26	999.34	999.51	998.80	1000.51
2	998.44	997.90	999.95	999.18	999.82
3	999.18	995.54	996.78	997.86	998.75
4	997.70	995.00	997.46	999.02	995.90
5	996.41	995.60	994.52	996.09	995.71
Normalized	1	2	3	4	5
1	0.791	0.805	0.834	0.715	1.000
2	0.654	0.565	0.906	0.777	0.885
3	0.777	0.171	0.378	0.557	0.706
4	0.530	0.080	0.492	0.752	0.231
5	0.316	0.181	0.000	0.263	0.200

Table 10.2 Initial parameters and final convolution layer values for the example of CNN in Example 10.2

		W0	W1	
	C1	−3.23	1.73	
	C2	3.03	−1.76	
Initial parameters	F	1	2	3
	1	−1.18	−2.25	−1.63
	2	0.38	0.01	0.12
	3	2.02	2.02	1.47
Conv. layer	C(a, b)	1	2	3
	1	−0.688	−0.981	−0.416
	2	−0.657	−0.876	−0.628
	3	−0.305	−0.574	−1.019

for true line. Note that if it is not northerly wind, the one hot encoding should be [0, 1]. See Eq. (4.18) for detail.

Note that the parameters of the current CNN model are initially generated as shown in the upper part of Table 10.2.

Solve the following:

(1) Normalize the MSLP data in Table 10.1 with the minmax scaler in Eq. (3.1).
(2) Estimate the prediction output value with the normalized input data.

(3) Train the CNN, including the filter values and weights of the fully connected layer (w1 and w0).

Solution:

(1) Normalization can be done with the minmax scaler in Eq. (3.1). For example,

$$x_i = \frac{x_i^{org} - x_{min}^{org}}{x_{max}^{org} - x_{min}^{org}} = \frac{999.26 - 993.48}{1003.11 - 993.48} = 0.60$$

At the bottom of Table 10.1, the normalized data is shown.

(2) The prediction output with the normalized data can be done with Eq. (10.14).

① Start with estimating the convolution layer as

$$C(1,1) = \sum_{a=1}^{N_{FH}} \sum_{b=1}^{N_{FW}} F_{a,b} I_{a,b} = 0.6 \times -1.18 + 0.609 \times -2.25 + \cdots + 0.343 \times 1.47 = -0.69$$

The estimated convolution values can be seen in the lower part of Table 10.2.

② MP is the maximum among the convolution layers (3, 3) as

$$MP = \max_{l \in N_{CH}, m \in N_{CW}} C(l, m) = -0.305$$

③ The fully connected layer is

$$\begin{bmatrix} z^1 = w_{11} MP_i + w_{01} = 1.73(-0.305) - 3.23 = -3.752 \\ z^2 = w_{12} MP + w_{02} = -1.76(-0.305) + 3.03 = 3.566 \end{bmatrix}$$

④ The final Softmax layer value is

$$y_p^1 = \frac{\exp(z_i^c)}{\sum_{c'=1}^{N_{sm}} \exp(z_i^{c'})} = \frac{\exp(-3.752)}{\exp(-3.752) + \exp(3.566)} = 0.00066$$

Also, $y_p^2 = 0.99934$

Note that the final output prediction value $\mathrm{y_p}$ is the probability that is comparable to the indication whether northerly or not.

(3) Training the CNN parameters starts with the estimating of yp as shown above.

① The derivative for the error (CEE) and Softmax layer can be easily estimated as

$$\left(y_p^1 - y_o^1\right) = 0.00066 - 1 = 0.99934$$

$$\left(y_p^2 - y_o^2\right) = 0.99934 - 0 = 0.99934$$

② For the parameters of the fully connected layer,

$$\frac{\partial E_i}{\partial w_{11}} = \frac{\partial E_i}{\partial y_p^1}\frac{\partial y_p^1}{\partial z^1}\frac{\partial z^1}{\partial w_{11}} = \left(y_p^1 - y_o^1\right)MP = 0.99934 \times (-0.305) = -0.305$$

$$\frac{\partial E_i}{\partial w_{01}} = \frac{\partial E_i}{\partial y_p^1}\frac{\partial y_p^1}{\partial z^1}\frac{\partial z^1}{\partial w_{01}} = \left(y_p^1 - y_o^1\right) = 0.99934$$

$\frac{\partial E_i}{\partial w_{12}} = -0.305$ and $\frac{\partial E_i}{\partial w_{02}} = 0.99934$

③ For the filter values

$$\frac{\partial E_i}{\partial F_{1,1}} = \sum_{c=1}^{2}\frac{\partial E_i}{\partial y_p^c}\frac{\partial y_p^c}{\partial z^c}\frac{\partial z^c}{\partial MP}\frac{\partial MP}{\partial C}\frac{\partial C}{\partial F_{a,b}} = \sum_{c=1}^{2}\left(y_p^c - y_o^c\right)w_{1c}\left(I_{l+a-1,m+b-1}\, for \text{ max } C(l,m)\right)$$
$$= 0.99934 \times 1.73 \times 0.591 + 0.99934 \times (-1.76) \times 0.591 = -0.0177$$

Note that 0.591 is chosen for $I_{l+a-1,m+b-1}$ since the corresponding value $I_{l+a-1,m+b-1}$ at max $C(l,m) = -0.305$ is $I_{3,1}$=0.591 as l=3, m=1 and $l + a - 1 = 3 + 1 - 1 = 3$ and $m + b - 1 = 1 + 1 - 1 = 1$.

④ The parameters are updated by summing all the errors E_i, i=1,…,N as in Eqs. (10.28) and (10.29) with the gradient descent algorithm.

[Python 10.1] Program the CNN in [Example 10.2]

```
import numpy as np
import pandas as pd
import matplotlib.pyplot as plt
from mpl_toolkits import mplot3d
import Functions_Ex9_CNN as Fcnn
#Simulated Data
TotLeng=200
HalfTL=100
XsimData=np.zeros((TotLeng,25))
for ist in range(5):
    XsimData[:HalfTL,ist*5:(ist+1)*5]=(1000-ist)+np.random.randn(HalfTL,5)
XsimData[HalfTL:200,:]=np.random.randn(HalfTL,25)+1000
Yobs_set=np.zeros(TotLeng)
Yobs_set[0:HalfTL]=1
YobsRe=np.zeros((TotLeng,2))
YobsRe[:,0]=Yobs_set
YobsRe[HalfTL:,1]=1
#
NHconv=5
NWcon=5
# Data transformation
ind_perm=np.random.choice(TotLeng,TotLeng)
Xobs_org=XsimData[ind_perm,:]
Xobs_tf,maxXobs,minXobs=Fcnn.Transf_MaxMin(Xobs_org)
Yobs_org=YobsRe[ind_perm,:]
Ntrain=100
Xobs_train=Xobs_tf[:100,:]
Xobs_valid=Xobs_tf[100:,:]
Yobs_train=Yobs_org[:100,:]
Yobs_valid=Yobs_org[100:,:]

###########################
# Parameter Initialization
Finit=np.random.rand(9)*0.1 #np.array([[0,1,0,0,1,0,0,1,0]],float)
Winit=np.random.randn(2,2)*0.1
F=Finit
W=Winit
NFh=3

Ypre,MP,indMP=Fcnn.CNN_forward(Xobs_train,F,W,NHconv,NWcon,NFh)
```

[Python 10.1] *Continued*

```
### Parameter Learning
count=0
Ncount=1000
Err_SE=np.zeros(Ncount)
Err_CEE=np.zeros(Ncount)
Learning_rate=0.001
```

```
while count<Ncount:
   W,F=Fcnn.update_para_CNN(Ypre,Yobs_train,Xobs_train,MP,indMP,F,W,Learning_rate,NHconv,NFh)
   Ypre,MP,indMP=Fcnn.CNN_forward(Xobs_train,F,W,NHconv,NWcon,NFh)
   Err_CEE[count]=Fcnn.ErrorX_CEE(Ypre,Yobs_train)
   Err_SE[count]=Fcnn.ErrorX_SE(Ypre,Yobs_train)
   count+=1

### Checking
plt.plot(Err_CEE,label='CEE')
plt.plot(Err_SE,label='SE')
plt.xlabel('Iteration')
plt.ylabel('Err')
plt.legend()
plt.show()
```

```
Ypre_train,MP,indMP =Fcnn.CNN_forward(Xobs_train,F,W,NHconv,NWcon,NFh)
Err_CEE_train=Fcnn.ErrorX_CEE(Ypre,Yobs_train)
Err_SE_train=Fcnn.ErrorX_SE(Ypre,Yobs_train)

Ypre_valid,MP,indMP =Fcnn.CNN_forward(Xobs_valid,F,W,NHconv,NWcon,NFh)
Err_CEE_valid=Fcnn.ErrorX_CEE(Ypre_valid,Yobs_valid)
Err_SE_valid=Fcnn.ErrorX_SE(Ypre_valid,Yobs_valid)


plt.plot(Yobs_valid[:,0],Ypre_valid[:,0],'.',color='red')
plt.plot(Yobs_valid[:,1],Ypre_valid[:,1],'.',color='blue')
plt.show()
```

```
[Python 10.1] Continue: Functions_Ex9_CNN -1
import numpy as np
import pandas as pd
import matplotlib.pyplot as plt
from mpl_toolkits import mplot3d
```

```
def EstConvolution1(Y,F,NF):
   Nih=len(Y)
   Niw=len(Y)
   Nch=Nih-NF+1
   Ncw=Niw-NF+1
   C=np.zeros([Nch,Ncw])
   Fre=F.reshape((NF,NF))
   for l in range(Nch):
      for m in range(Ncw):
         C[l,m] = sum(sum(Fre*Y[l:l+NF,m:m+NF]))
   return C.reshape(Nch*Ncw)
```

```
def Transf_MaxMin(XX):
    return (XX-XX.min())/(XX.max()-XX.min()), XX.max(),XX.min()

def TransfBack_MaxMin(XX,maxX,minX):
    return XX*(maxX-minX)+minX

def EstAllConv(Xall,F,NHconv,NWcon,NFh):
    Nd=len(Xall)
    Nch=NHconv-NFh+1
    Ncw=NWcon-NFh+1

    Call=np.zeros([Nd,Nch*Ncw])

    for iy in range(Nd):
        Y=Xall[iy,:].reshape((NHconv,NWcon))
        Call[iy,:]=EstConvolution1(Y,F,NFh)

    return Call
```

[Python 10.1] *Continue:* Functions_Ex9_CNN -2

```
def EstPooling(Xd,Ptype):
    nD=len(Xd)
    yout=np.zeros(nD)
    indS=np.zeros(nD,int)
    if Ptype==0:#Max pooling
        for idx in range(nD):
            yout[idx]=Xd[idx,:].max(0)
             indX=np.where(Xd[idx,:]==yout[idx])
            indS[idx]=indX[0]
    elif Ptype==1:
        yout=Xd.mean(0)
    else:
        yout=0
    return yout, indS

def CNN_forward(Xobs,F,W,NHconv,NWcon,NFh):
    Cdata=EstAllConv(Xobs,F,NHconv,NWcon,NFh)
    MP, indMP=EstPooling(Cdata,0)
    Z=np.zeros((len(Xobs),2))
    Z[:,0]=W[0,0]*MP+W[0,1]
    Z[:,1]=W[1,0]*MP+W[1,1]
    Zexp=np.exp(Z)
    Y=np.zeros((len(MP),2))
    Y[:,0]=Zexp[:,0]/np.sum(Zexp,1)
    Y[:,1]=Zexp[:,1]/np.sum(Zexp,1)
    return Y, MP, indMP

def dE_dW(Ypred,YobsRe,MP):
    dW=np.zeros((2,2))
    for ir in range(2):
        dW[ir,0]=-np.sum((YobsRe[:,ir]-Ypred[:,ir])*MP)
        dW[ir,1]=-np.sum((YobsRe[:,ir]-Ypred[:,ir]))
    return dW
```

```
[Python 10.1] Continue: Functions_Ex9_CNN -3

def dE_dF(Ypred,YobsRe,MP,W,Xobs,MP_ind,NHconv,NFh):
  X_dF=np.zeros(NFh**2)
  nD=len(Xobs)
  FF0=np.zeros((nD,NFh**2))
  FF1=np.zeros((nD,NFh**2))
  for ist in range(len(Xobs)):
    l=np.int(np.floor(MP_ind[ist]/NFh))
    m=np.remainder(MP_ind[ist],NFh)
    for ir in range(NFh):
      X_dF[ir*NFh:(ir+1)*NFh]=Xobs[ist,(ir*NHconv)+l*NHconv+m:(ir*NHconv)+l*NHconv+m+NFh]

    FF0[ist,:]=(YobsRe[ist,0]-Ypred[ist,0])*W[0,0]*X_dF
    FF1[ist,:]=(YobsRe[ist,1]-Ypred[ist,1])*W[1,0]*X_dF
                              dF=-np.sum(FF0,0)-np.sum(FF1,0)

  return dF


def update_para_CNN(Ypred,YobsRe,Xobs,MP,MP_ind,Fnow,Wnow,Learning_rate,NHconv,NFh):
  Wnow -= Learning_rate*dE_dW(Ypred,YobsRe,MP)
  Fnow -= Learning_rate*dE_dF(Ypred,YobsRe,MP,Wnow,Xobs,MP_ind,NHconv,NFh)
  return Wnow, Fnow

def ErrorX_CEE(Ypred,Yobs):
  return 0.5*np.sum(-Yobs*np.log(Ypred))

def ErrorX_SE(Ypred,Yobs):
  return 0.5*np.sum((Yobs-Ypred)**2)
```

10.3 Exercises

1. Estimate the output of CNN for the scaled data of Table 10.1 employing the following filters.

   (a) Filter 1

0.5	0.5	0.5
0	0	0
1	1	1

   (b) Filter 2

1	2	3
1	2	3
1	2	3

2. Estimate the output of CNN for the scaled data of Table 10.1 employing the filters in Exercise 1
   (a) Include padding with zero values
   (b) Include padding with one values
   (c) Include padding (zeros) and stride of (2, 2)
3. Train the CNN model (backpropagation) in Fig. 10.2 with standardizing the original dataset of Table 10.1 instead of minmax scaler. (Fig. 10.1)

Chapter 11
Tensorflow and Keras Programming for Deep Learning

Abstract Tensorflow is an end-to-end open-source platform for machine learning containing a comprehensive, flexible ecosystem of tools, libraries, and community resources (https://www.tensorflow.org/). It provides multiple levels of abstractions to choose the right one. The high-level Keras API can be used to build and train models by easily getting started with Tensorflow. Keras allows employing Tensorflow without losing its flexibility and capability. In the following, two applications (i.e., temporal and spatial deep learning) are presented to illustrate how to use Keras with python.

11.1 Basic Keras Modeling

In order to learn an ANN model in Python with Keras, the related functions must be imported as described below:

```
from keras.models import Sequential
from keras.layers import Dense, Activation
from keras.optimizers import adam
from sklearn.metrics import mean_squared_error
```

The "Sequential" function must be imported to define the deep learning model sequentially in advance. Dense and Activation is the function to define the ANN and its activation function. The "adam" is also imported to train the employed model as in Sect. 6.4.

One can easily apply the simple neural network as the sequential setup modeling for the input data as "Xi" and output data "yo" with the predefined hidden unit as NumHid = 3 (see Fig. 4.1). The model can be constructed and trained as the following code:

T. Lee et al., *Deep Learning for Hydrometeorology and Environmental Science*, Water Science and Technology Library 99,
https://doi.org/10.1007/978-3-030-64777-3_11

```
model = Sequential()

model.add(Dense(NumHid))

model.add(Activation('tanh'))

model.add(Dense(1))

model.add(Activation('linear'))

model.compile(loss='mean_squared_error', optimizer='adam')

history=model.fit(X_train, Y_train, epochs=1000,batch_size=100,validation_split=0.33)
```

In this code, the model is defined as "model = Sequential()." Then, in this model, the structure is added sequentially as left to right of Fig. 4.1. The three hidden neurons are added at first as "model.add(Dense(NumHid))" and it is activated with tanh function as "model.add(Activation('tanh'))." Then, the hidden state outflows into one output with linear activation function as "model.add(Dense(1))" and "model.add(Activation('linear'))."

After that, a little setting must be made and then the data must be fitted as follows.

```
model.compile(loss='mean_squared_error', optimizer='adam')

history=model.fit(Xi, yo, epochs=30, validation_split=0.33,verbose=False)
```

Here, the mean squared error function is used as the loss function and the optimizer "adam" is used. By setting the model, the following line executes to train the model as the compile setup with 30 epochs and the validation split portion is 33%. Note that "verbose=False" indicates that its execution will not be displayed in detail while running the program. The data dimension for input and output is automatically detected by the input. Therefore, the input and output data (i.e., Xi and yo) must be appropriately arranged.

Here, one example with the Pacific Decadal Oscillation (PDO) is presented to describe how to apply Keras to a neural network model.

```
import pandas as pd
CI=pd.read_csv('PDO_2017_CSV.csv',header=None).values

#1900,0.04,1.32,0.49,0.35,0.77,0.65,0.95,0.14,-0.24,0.23,-0.44,1.19

dataset = np.asarray(CI[:,1:-1]).reshape(-1) # rearrange the dataset into one column
```

The PDO dataset is uploaded as an example with the excel file and it is monthly data and the dataset is with the year in the first column and the annual mean in the last column. The dataset is read by "pandas" function as "CI=pd.read_csv('PDO_2017_CSV.csv',header=None).values" and it is rearranged with just one column as "CI1=np.asarray(CI[:,1:−1]).reshape(−1)." Note that reshape(−1) can be used without knowing the size of the original dataset (here, CI) with just one column.

```
num_train=int(0.8*len(dataset)) #80% of the record will be used for train and validation

# Minmax scaling

minX = dataset[:num_train].min(axis=0)

maxX = dataset[:num_train].max(axis=0)

difNX=maxX-minX

dataset_stz = (dataset-minX)/difNX
```

The dataset is scaled (or normalized) at first by minmax scaling. Its min and max are defined with the training dataset because the fitted model with the training data is finally employed. Here, 80% of the record is used for training and validation, and the other 20% is used for testing.

In addition, the number of inputs is two as in the example of Fig. 4.1. Therefore, two step backward dataset is used as input and the dataset must be assigned accordingly as the following code.

```
look_back=2 # look_back indicates the number of inputs

dataset_all=np.zeros((len(dataset_stz)-look_back,look_back+1))

for it in range(look_back+1):

   dataset_all[:,it]=dataset_stz[it:len(dataset)-(look_back)+it]
```

Finally, the training and testing dataset is assigned as follows.

```
X_train=dataset_all[0:num_train,:-1]

Y_train=dataset_all[0:num_train,-1]

X_test=dataset_all[num_train:,:-1]

Y_test=dataset_all[num_train:,-1]
```

Note that the validation dataset is not defined here since the validation set will be automatically assigned in the fitting process with "validation_split."

```
[Python 11.1] Program the simple ANN with Keras in Figure 4.1
import pandas as pd
from sklearn.metrics import mean_squared_error
from keras.models import Sequential
from keras.layers import Dense, Activation, LSTM
from keras.optimizers import adam

CI=pd.read_csv('PDO_2017_CSV.csv',header=None).values
CI1 = np.asarray(CI[:,1:-1]).reshape(-1)
dataset=CI1
num_train=int(0.8*len(dataset))
minX = dataset[:num_train].min(axis=0)
maxX = dataset[:num_train].max(axis=0)
difNX=maxX-minX
dataset_stz = (dataset-minX)/difNX

data_train=dataset_stz[0:num_train]
data_test=dataset_stz[num_train:]

look_back=2
dataset_all=np.zeros((len(dataset_stz)-look_back,look_back+1))

for it in range(look_back+1):
   dataset_all[:,it]=dataset_stz[it:len(dataset)-(look_back)+it]

X_train=dataset_all[0:num_train,:-1]
Y_train=dataset_all[0:num_train,-1]
X_test=dataset_all[num_train:,:-1]
Y_test=dataset_all[num_train:,-1]

## Modeling
model = Sequential()
model.add(Dense(NumHid))
model.add(Activation('tanh'))
model.add(Dense(1))
model.add(Activation('linear'))
model.compile(loss='mean_squared_error', optimizer='adam')
history=model.fit(X_train, Y_train, epochs=1000,batch_size=100,validation_split=0.33)

Y_pred_train=model.predict(X_train)*difNX + minX
Y_pred_test=model.predict(X_test)*difNX + minX
```

The whole code is shown as follows.

11.2 Temporal Deep Learning (LSTM and GRU)

To apply LSTM with Keras is very simple and direct. It is almost identical to the ANN model except that the LSTM layer is only replaced instead of Dense and tanh layer. The LSTM modeling of Fig. 9.2 by creating a function can be coded as follows:

```
def fit_lstm(train, batch_size, nb_epoch, neurons):
    X, y = create_dataset(train, look_back)
    X = X.reshape(X.shape[0], X.shape[1],1 )

    model = Sequential()
    model.add(LSTM(neurons[0],input_shape=( X.shape[1], X.shape[2])))
    model.add(Dense(1))
    model.add(Activation('linear'))
    model.compile(loss='mean_squared_error', optimizer='adam')
    history=model.fit(X, y, epochs=nb_epoch,batch_size=batch_size,validation_split=0.33)
    return model, history
```

Here, the LSTM layer should include the number of hidden units (or called neurons) and input shape. The input_shape takes two-dimensional numbers. The first one is the number of time steps to consider and the second one is the number of output units (or the number of features). Note that the sample size is not included. Therefore, the input data (**X**) dimension must match the input_shape as (sample size (N) $\times$ number of time steps $\times$ number of features). That is why the input_shape is defined as input_shape = (X.shape[1], X.shape[2]). The output unit (here X.shape[2]) should coincide with the number of output dimension, as the output data dimension is sample size (N) $\times$ number of output units (here, X.shape[2]).

The data can be preprocessed for input and output data by defining a function as follows:

```
def create_dataset(dataset, look_back=1):
    dataX, dataY = [], []
    for i in range(len(dataset)-look_back-1):
        a = dataset[i:(i+look_back)]
        dataX.append(a)
        dataY.append(dataset[i + look_back])
    return np.array(dataX), np.array(dataY)
```

In this code, the look_back indicates the time steps to consider. Also, the X dataset must be rearranged to meet three-dimensional format for the LSTM as.

```
X, y = create_dataset(dataset, look_back)

X = X.reshape(X.shape[0], X.shape[1],1 )
```

Furthermore, one can add the additional inputs as stateful and return_sequences. When one wants to use the stacked LSTM (more than one LSTM layer), return_sequences=True can be used (default is False). If stateful=True. Then, hidden and cell states are inherited at sequence batches. Note that the hidden and cell states start with zero values at each batch with stateful=False.

In the following, the LSTM model is fitted to the Pacific Decadal Oscillation (PDO) monthly data to forecast the future PDO index downloaded from https://www.ncdc.noaa.gov/teleconnections/pdo/ with the csv format from year and 12-monthly values and annual average. PDO, as Pacific Decadal Oscillation (PDO), is one of the most representative climate indices, a long-lived El Nino-like pattern of Pacific climate variability (Mantua et al. 1997). This index has been proven to be teleconnected with many hydrometeorological variables in the literature (Nigam et al. 1999).

One month ahead prediction has been performed with the LSTM model with 4 hidden units. Note that three performance measurements are applied as RMSE, MAE, and r^2 with the original domain for minmax scaler (see Eq. 3.3).

The whole code is shown as [Python 11.2] and its result is shown at Fig. 11.1. The result seems satisfactory to predict one month ahead PDO (top panel) with the LSTM model. The training seems reasonable according to the loss function of training and validation batches.

[Python 11.2] Program the LSTM with Keras in Figure 4.1

```
import pandas as pd
from sklearn.metrics import mean_squared_error, mean_absolute_error, r2_score
from keras.models import Sequential
from keras.layers import Dense, Activation, LSTM
from keras.optimizers import adam
from math import sqrt
import numpy as np

def create_dataset(dataset, look_back=1):
        dataX, dataY = [], []
        for i in range(len(dataset)-look_back-1):
                a = dataset[i:(i+look_back)]
                dataX.append(a)
                dataY.append(dataset[i + look_back])
        return np.array(dataX), np.array(dataY)

# fit an LSTM network to training data
def fit_lstm(train, batch_size, nb_epoch, neurons):
    X, y = create_dataset(train, look_back)
    X = X.reshape(X.shape[0], X.shape[1],1 )
```

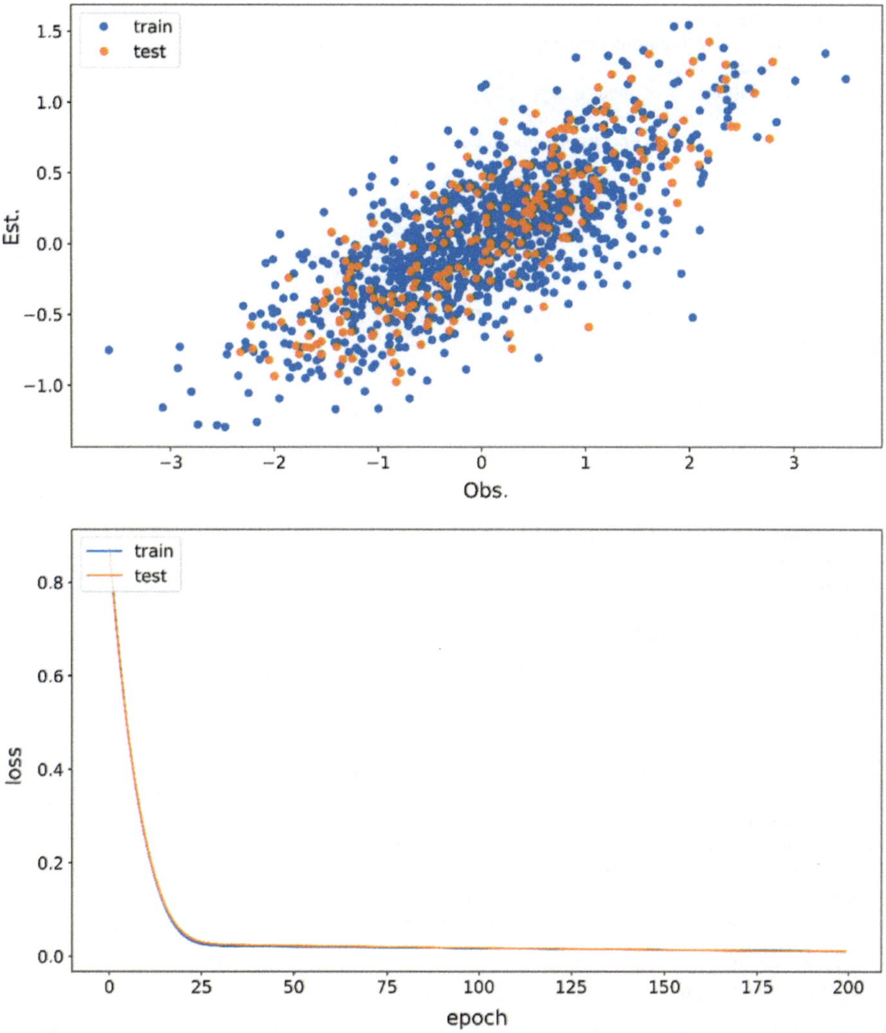

Fig. 11.1 Results of LSTM for one step ahead prediction of the monthly PDO index. The scatter plot (top panel) between the observed and the estimated with the fitted LSTM model is shown and the loss function (here, RMSE) along with the epoch (bottom panel)

```
        model = Sequential()
        model.add(LSTM(neurons[0],input_shape=( X.shape[1], X.shape[2])))
        model.add(Dense(1))
        model.add(Activation('linear'))
        model.compile(loss='mean_squared_error', optimizer='adam')
        history=model.fit(X, y, epochs=nb_epoch,batch_size=batch_size,validation_split=0.33)

    return model, history
```

```
[Python 11.2] Continue
CI=pd.read_csv('PDO_2017_CSV.csv',header=None).values
CI1 = np.asarray(CI[:,1:-1]).reshape(-1)
dataset=CI1

neurons=[4]
nb_epoch=200
look_back=12

num_train=int(0.8*len(dataset))
minX = dataset[:num_train].min(axis=0)
maxX = dataset[:num_train].max(axis=0)
difNX=maxX-minX

dataset_stz = (dataset-minX)/difNX
```

```
data_train=dataset_stz[0:num_train]
data_test=dataset_stz[num_train:]

# fit the model
lstm_model,hisT = fit_lstm(data_train, 100, nb_epoch, neurons)

#forecast test data
X_train, y_train = create_dataset(data_train, look_back)
X_test, y_test = create_dataset(data_test, look_back)
X_test = X_test.reshape(X_test.shape[0], X_test.shape[1],1 )
X_train = X_train.reshape(X_train.shape[0], X_train.shape[1],1 )
```

```
y_pred_test=lstm_model.predict(X_test)
y_pred_train=lstm_model.predict(X_train)

#Back-transformation to original domain
y_pred_test_BT=y_pred_test*difNX+minX
y_true_test_BT=y_test*difNX+minX

y_pred_train_BT=y_pred_train*difNX+minX
y_true_train_BT=y_train*difNX+minx

# Performance measurements
rmse_test = sqrt(mean_squared_error(y_true_test_BT, y_pred_test_BT))
MAE_test = mean_absolute_error(y_true_test_BT, y_pred_test_BT)
r2 = r2_score(y_true_test_BT, y_pred_test_BT)
```

The GRU model can be easily applied by replacing the LSTM layer with the GRU layer as.

```
model = Sequential()

model.add(GRU(neurons[0],input_shape=( X.shape[1], X.shape[2])))

model.add(Dense(1))

model.add(Activation('linear'))

model.compile(loss='mean_squared_error', optimizer='adam')

history=model.fit(X, y, epochs=nb_epoch,batch_size=batch_size,validation_split=0.33)
```

11.3 Spatial Deep Learning (CNN)

In Keras, CNN is modeled with Conv2D in a sequential model. The main inputs for the Conv2D are the number of filters, kernel size, input shape, and activation function. As an example, in Fig. 10.6, the number of filters is one. The kernel size is $N_{FH} \times N_{FC,}$ as in Eq. (10.4). The input shape must be three dimensional as the number of convolution for row and column (see N_{CW} and N_{CH} in Eq. (10.5)). This is the dimension of each two-dimensional input data, excluding the sample size.

The example model in Fig. 10.6 can be defined as follows. The Max Pooling algorithm in Eq. (10.10) can be done with the MaxPooling2D layer as “model.add(MaxPooling2D(pool_size=3)).” The “Flatten()” layer must be added to arrange two dimension convolution data to one dimension data as model.add(Flatten()). The softmax activation function can be applied to connect the output as Eq. (10.16) by model.add(Dense(2,activation=“softmax”)). The model can be coded as follows.

```
model=Sequential()

model.add(Conv2D(1, kernel_size=(3, 3), input_shape=(NHconv,NWcon, 1), activation='relu'))

model.add(MaxPooling2D(pool_size=3))

model.add(Flatten())

model.add(Dense(2,activation="softmax"))

model.compile(loss='categorical_crossentropy',optimizer='adam',metrics=['accuracy'])

history=model.fit(X_train,Y_train,validation_split=0.3, epochs=3500,batch_size=50,verbose=0)
```

The whole code is in [Python 11.3]. The result for the CNN model in Fig. 10.6 is shown in Fig. 11.2. The output of the final CNN consists of the probability as [0, 1]. The result indicates that the applied CNN model is well fitted to the simulated dataset (see the top panel of Fig. 10.6) and its fitting process seems appropriate by smoothly decreasing the loss function defined here as categorical cross-entropy in Eq. (4.14) (see the bottom panel of Fig. 10.6).

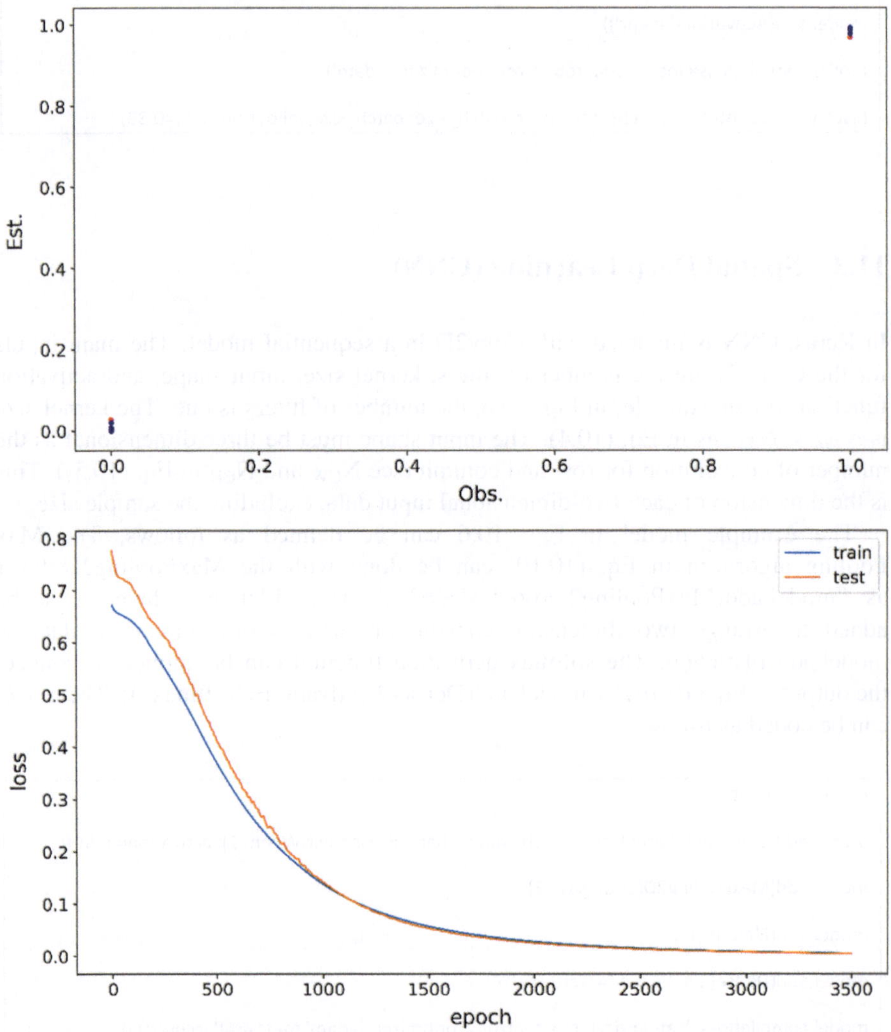

Fig. 11.2 Results of CNN for the example in Fig. 10.6. The scatter plot (top panel) between the observed and the estimated with the fitted LSTM model is shown and the loss function (here, RMSE) along with the epoch (bottom panel)

[Python 11.3] *CNN modeling in* [Example 10.2] *with Keras*

```
from keras.models import Sequential
from keras.layers import Flatten, Conv2D, MaxPooling2D
from keras.utils import np_utils
from keras.callbacks import EarlyStopping
import matplotlib.pyplot as plt
import numpy as np

# Data Simulation
TotLeng=200
HalfTL=100
XsimData=np.zeros((TotLeng,25))
for ist in range(5):
    XsimData[:HalfTL,ist*5:(ist+1)*5]=(1000-ist)+np.random.randn(HalfTL,5)
XsimData[HalfTL:200,:]=np.random.randn(HalfTL,25)+1000
Yobs_set=np.zeros(TotLeng)
Yobs_set[0:HalfTL]=1
YobsRe=np.zeros((TotLeng,2))
YobsRe[:,0]=Yobs_set
YobsRe[HalfTL:,1]=1
#
NHconv=5
NWcon=5
# Data preparation
ind_perm=np.random.choice(TotLeng,TotLeng)
Xobs_org=XsimData[ind_perm,:]
Xobs_tf=(Xobs_org-Xobs_org.min())/(Xobs_org.max()-Xobs_org.min())
Yobs_org=YobsRe[ind_perm,:]
```

```
Ntrain=100
Xobs_train=Xobs_tf[:100,:]
Xobs_valid=Xobs_tf[100:,:]
Yobs_train=Yobs_org[:100,:]
Yobs_valid=Yobs_org[100:,:]

X_train=Xobs_train.reshape(Xobs_train.shape[0],5,5,1)
Y_train=Yobs_train[:,0]
Y_train=np_utils.to_categorical(Y_train,2)
X_valid=Xobs_valid.reshape(Xobs_valid.shape[0],5,5,1)
Y_valid=Yobs_valid[:,0]
Y_valid=np_utils.to_categorical(Y_valid,2)
```

[Python 11.3] *Continue*

```
X_train=Xobs_train.reshape(Xobs_train.shape[0],5,5,1)
Y_train=Yobs_train[:,0]
Y_train=np_utils.to_categorical(Y_train,2)

X_valid=Xobs_valid.reshape(Xobs_valid.shape[0],5,5,1)
Y_valid=Yobs_valid[:,0]
Y_valid=np_utils.to_categorical(Y_valid,2)
```

```
# CNN model fitting
model=Sequential()
model.add(Conv2D(1, kernel_size=(3, 3), input_shape=(NHconv,NWcon, 1), activation='relu'))
model.add(MaxPooling2D(pool_size=3))
model.add(Flatten())
model.add(Dense(2,activation="softmax"))
model.compile(loss='categorical_crossentropy',optimizer='adam',metrics=['accuracy'])
history=model.fit(X_train,Y_train,validation_split=0.3, epochs=3500,batch_size=50,verbose=0)

## For early stopping
#early_stopping_callback=EarlyStopping(monitor='val_loss',patience=50)
#history=model.fit(X_train,Y_train,validation_split=0.3, epochs=10000,batch_size=50,
#            verbose=1,callbacks=[early_stopping_callback]);

y_pre_tr=model.predict(X_train)
y_pre_val=model.predict(X_valid)

plt.plot(Y_train[:,0],y_pre_tr[:,0],'.',color='red')
plt.plot(Y_valid[:,0],y_pre_val[:,0],'.',color='blue')
plt.show()

plt.plot(history.history['loss'])
plt.plot(history.history['val_loss'])
plt.legend(['train', 'test'], loc='upper right')
plt.show()
```

11.4 Exercises

1. Program a neural network model with the PDO data with Keras. Set the number of hidden units as 10 and use the sigmoid function for the activation function of the hidden layer and linear function for the output layer. Assume $\mathbf{X} = [y_{t-1}, y_{t-2}, y_{t-3}]$, $\mathbf{Y} = [y_t]$.
2. Program the LSTM model for the PDO data with 10 hidden units and $\mathbf{X} = [y_{t-1}, y_{t-2}, y_{t-3}]$, $\mathbf{Y} = [y_t]$.
3. Code the CNN shown in Fig. 10.6, but set the filter dimension as 4-by-4 instead of 3-by-3.

References

Mantua NJ, Hare SR, Zhang Y, Wallace JM, Francis RC (1997) A Pacific interdecadal climate oscillation with impacts on salmon production. Bull Am Meteor Soc 78(6):1069–1079

Nigam S, Barlow M, Berbery EH (1999) Analysis links pacific decadal variability to drought and streamflow in united states. Eos 80(51):621–622+625

Chapter 12
Hydrometeorological Applications of Deep Learning

Abstract Deep learning models have been applied for hydrometeorological applications. In this chapter, a few of them are explained. In the field of hydrometeorology, time-series deep learning models are mainly employed. In this chapter, the development procedure of a time series deep learning model for stochastic simulation producing a long sequence that mimics historical series is explained. Furthermore, the case study for daily maximum temperature with an LSTM model is presented.

12.1 Stochastic Simulation with LSTM

A long-term record of data is generated by stochastic time series models reproducing the statistical characteristics of time series data, not predicting specific values. Simulating a long-term record without the loss of original statistical characteristics is not an easy task due to the dissipation of information by the lead time. The statistical characteristics of hydrometeorological time series data are governed by a complex interaction of atmospheric and hydrologic phenomena. Conventional neural networks may not be able to reproduce the complexity in the data. Because deep neural networks have complex and deep hidden layers, they can account for more complex interaction. Therefore, the LSTM model was tested by Lee et al. (2020) with a stochastic simulation model as a deep neural network, and its performance was compared with traditional time series models such as AR and RNN. Here the work of Lee et al. (2020) is followed.

12.1.1 Mathematical Description for Stochastic Simulation with LSTM

Stochastic simulation with machine learning algorithms, such as RNN and LSTM, can be performed by setting the explanatory variables as the same variables but with time-lagged as $\mathbf{x}_t = [y_{t-1}, y_{t-2}, \ldots, y_{t-d}]$ as:

T. Lee et al., *Deep Learning for Hydrometeorology and Environmental Science*, Water Science and Technology Library 99,
https://doi.org/10.1007/978-3-030-64777-3_12

$$Y_t = f(\mathbf{x}_t) + \varepsilon_t \tag{12.1}$$

where $\mathbf{x}_t = [x_t^{\,1}, x_t^{\,2}, \ldots, x_t^{\,d}]$ is a set consisting of d number of predictor (or explanatory) variables at time t and ε_t is the independent white noise term regarded as $\varepsilon_t \sim N(0, \sigma_\varepsilon^{\,2})$, while Y_t is the predict and at time t. N and $\sigma_\varepsilon^{\,2}$ are the normal distribution and the variance of independent white noise term, respectively.

The functional relation, that is, $f(\mathbf{x}_t)$, between $\mathbf{x}_t$ and Y_t is nonlinearly created from the LSTMs. Input data sets for all employed models are preprocessed in advance. It is straightforward to simulate with the LSTM model after building its architecture and fitting it. The random numbers should be generated at each time step t from a normal distribution for ε_t and is multiplied with the error standard deviation σ_ε(i.e., $\varepsilon_t \sim N(0, \sigma_\varepsilon^{\,2})$ to preserve its variability. The error variance ($\sigma_\varepsilon^{\,2}$) can be estimated with the difference between the values predicted by machine learning algorithms and observed values.

The dataset for estimating the error variance must be obtained from the testing data without the training data since the real error for unknown input values is with the testing data. Otherwise, the variability of the simulated data is often underestimated.

12.1.2 Colorado Monthly Streamflow

Colorado River is one of the major river systems in North America (Lee et al. 2010), which is an indispensable source of water supply for seven states (i.e., Colorado, Utah, Arizona, New Mexico, Nevada, California, and Wyoming) in the western United States and Mexico. The Colorado River system has been divided into two large systems as the Upper and Lower Colorado River systems, where the division is at Lees Ferry. The system was analyzed by performing simulation studies using potential streamflow scenarios that may occur in the future.

For stochastic simulation, the monthly streamflow data from the Colorado River station below the Hoover Dam was used since the system was regulated on a monthly basis to assess model performance. Note that the historical data of the Colorado River has been extended up to 1906–2003 (Lee and Salas 2006).

12.1.3 Results of Colorado River

For the LSTM model, the monthly streamflow data were normalized for each month. To find the optimum number of hidden units, k-fold cross-validation (KfCV) was performed as in Sect. 7.2. Its result is shown in Fig. 12.1. The result indicates that the number of hidden units as four and five shows the best performance.

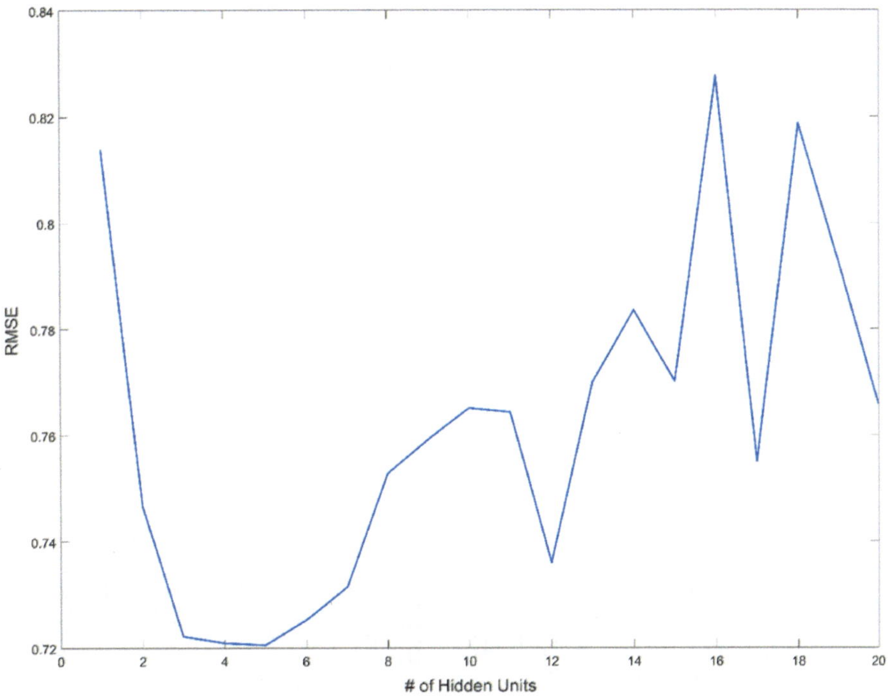

Fig. 12.1 K-fold cross-validation for the number of hidden units with monthly streamflow dataset from the Colorado River system below the Hoover Dam

The reproduction of the key statistics for different time scales, such as monthly and yearly, is critical in stochastic simulation. To check the performance, 100 sequences with the same record length were simulated from the LSTM stochastic model. The monthly key statistics of the simulated 100 sequences of monthly streamflow are presented in Fig. 12.2 with boxplots as well as observed data with the dash-dotted line with cross markers. Results show that the observed mean and standard deviation of monthly streamflow (top panels of Fig. 12.2) were well reproduced by the LSTM model, but skewness and lag-1 autocorrelation (middle panels of Fig. 12.2) were not much preserved.

The observed key statistics of the yearly series were well reproduced by the LSTM model, including the standard deviation and lag-1 correlation shown in Fig. 12.3. This indicates that the LSTM model reproduces the long-term variability embedded in the Colorado River system. The performance of the LSTM stochastic model is quite surprising that the LSTM model can reproduce these statistics at the yearly scale.

Reproduction of drought-related statistics is critical for stochastic simulation for managing the water resources of the Colorado River system. In Fig. 12.4, the consecutive length and sum for deficit and surplus as well as storage capacity were presented with the boxplots for data simulated by the LSTM model and the observed data

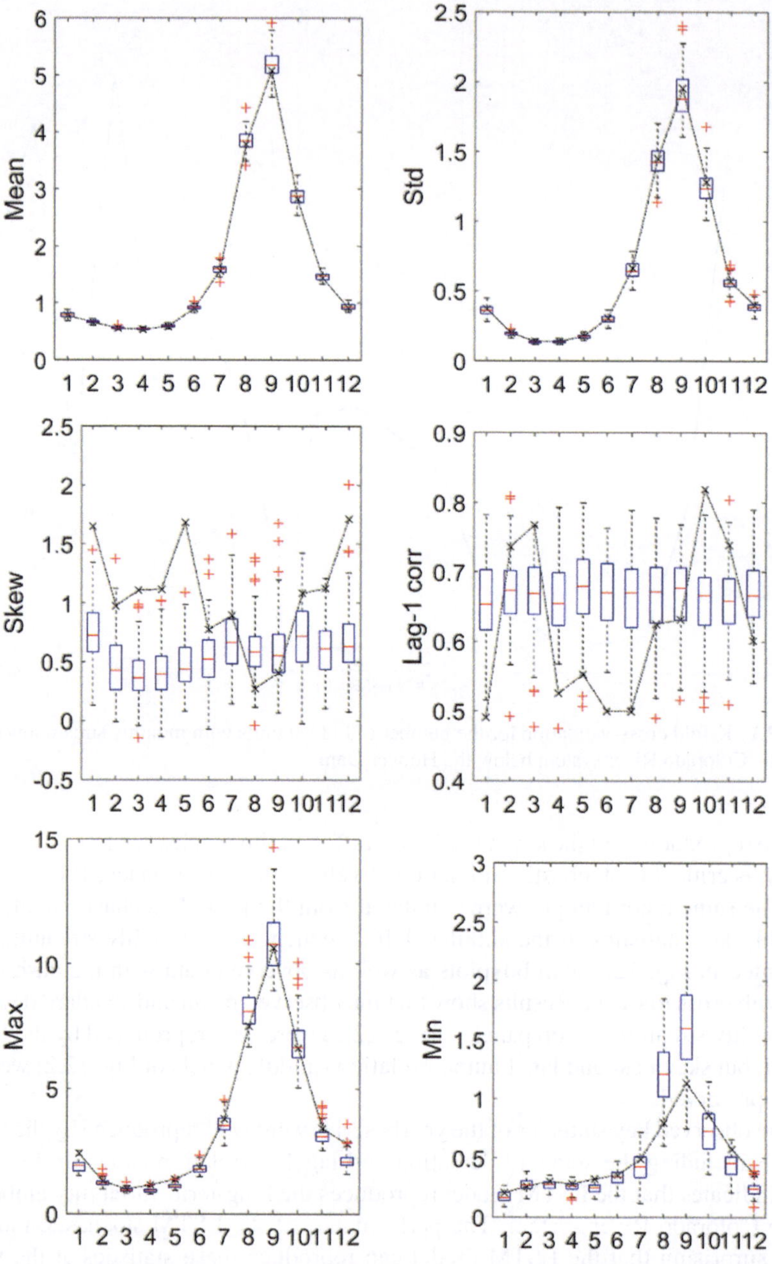

Fig. 12.2 Key statistics of the simulated data (boxplot) with the LSTM model and observed (dash-dotted line with x marker) for monthly streamflow ($10^9 m^3$) below the Hoover Dam station of the Colorado River system

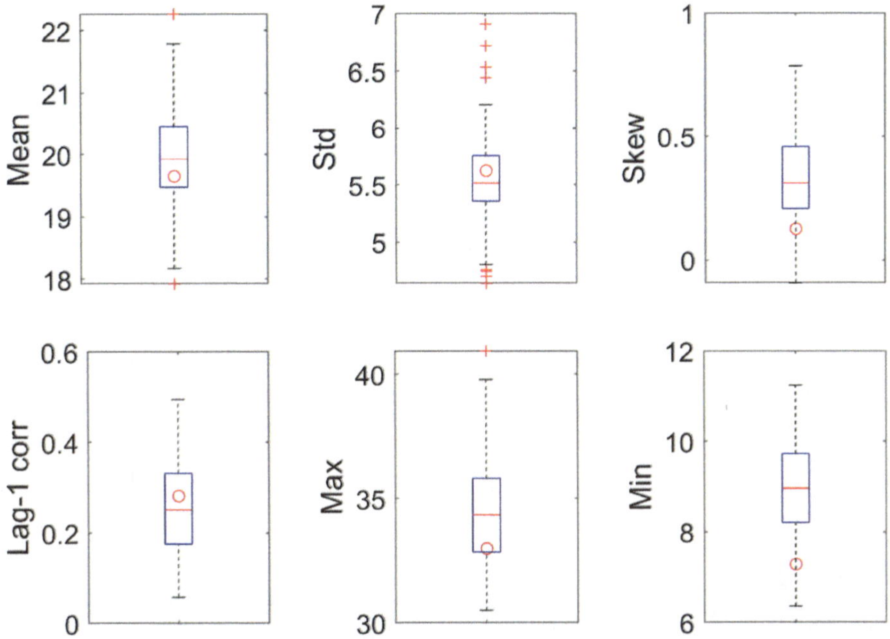

Fig. 12.3 Key statistics of the simulated data (boxplot) with the LSTM model and observed (circle) for the annual streamflow (10^9m^3) below the Hoover Dam station

(circle). The storage capacity was the minimum capacity that a reservoir could deliver a certain demand through the time period (Salas 1993). The LSTM model reasonably reproduced the storage capacity, while some biases were observed for other statistics in a reasonable manner.

12.1.4 Python Coding

In the python coding for the stochastic simulation of the Colorado River monthly streamflow, the LSTM model must be fitted to the dataset. The Keras code in [Python 11.2] can be used for the data preprocessing and LSTM model fitting as "def create_dataset(dataset, look_back=1)" and "def fit_lstm(train, batch_size, nb_epoch, neurons)."

The major coding part of stochastic simulation is (1) to estimate the error variance from the predicted values and (2) to add the random noise term as in Eq. (12.1) with the error variance estimated in step (1). The estimate of error variance as (1) can be done by.

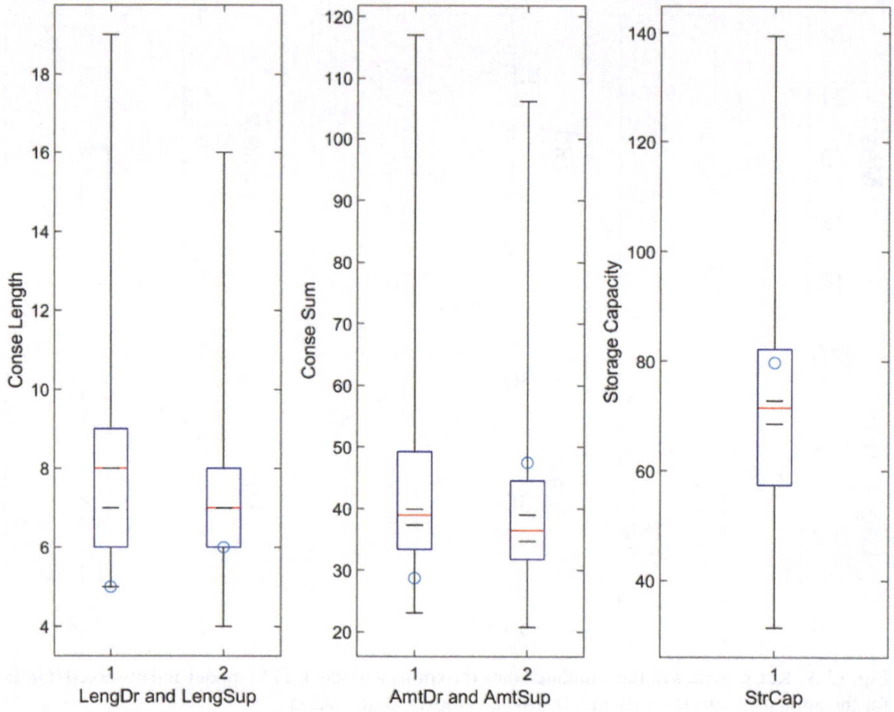

Fig. 12.4 Drought statistics of the simulated data (boxplot) with the LSTM model and observed (circle) for the monthly streamflow below the Hoover Dam station

```
lstm_model,hisT = fit_lstm(data_train, 100, nb_epoch, neurons)


X_train, y_train = create_dataset(data_train, look_back)

X_test, y_test = create_dataset(data_test, look_back)

X_test = X_test.reshape(X_test.shape[0], X_test.shape[1],1 )

X_train = X_train.reshape(X_train.shape[0], X_train.shape[1],1 )


y_pred_test=lstm_model.predict(X_test)


std_err=np.sqrt(np.mean((y_test-y_pred_test)**2))
```

Simulation can be done by adding the error term to the predicted value with the previous simulation term as.

```
        inpx=np.reshape(simX[is,:,:],(1,2,1))

        yt=lstm_model.predict(inpx)

        simX[is+1,1]=yt+np.random.normal(loc=0,scale=std_err,size=1)
```

The whole python coding can be seen as follows:

```
[Python 12.1] Python program for LSTM to forecast the daily temperature.
ColRiv=pd.read_csv('Colorado_MonStream.csv',header=0).values

ColRiv1 = np.asarray(ColRiv[:,1:-1]).reshape(-1)
dataset=ColRiv1

neurons=[4]
nb_epoch=200
look_back=2
#chia data thanh test, train, valid roi moi dung function create_data
Ndata=len(dataset)

num_train=int(0.8*len(dataset))
minX = dataset[:num_train].min(axis=0)
maxX = dataset[:num_train].max(axis=0)
difNX=maxX-minX
dataset_stz = (dataset-minX)/difNX

data_train=dataset_stz[0:num_train]
data_test=dataset_stz[num_train:]
# fit the model
lstm_model,hisT = fit_lstm(data_train, 100, nb_epoch, neurons)

X_train, y_train = create_dataset(data_train, look_back)
X_test, y_test = create_dataset(data_test, look_back)
X_test = X_test.reshape(X_test.shape[0], X_test.shape[1],1 )
X_train = X_train.reshape(X_train.shape[0], X_train.shape[1],1 )

y_pred_test=lstm_model.predict(X_test)
std_err=np.sqrt(np.mean((y_test-y_pred_test)**2))
```

```
[Python 12.1] Continue

N_it=100
N_sim=Ndata
Ntrain=len(X_train)
simZ=np.zeros([N_sim,N_it])

#perform the simulation
for it in range(N_it):
   simX = np.zeros([N_sim,look_back,1])
   simY = np.zeros([N_sim])
   t=random.randint(1,Ntrain)
   simX[0,:]=X_train[t,:]
   simY[0]=y_train[t]
   for is in range(N_sim-1):
```

```
        inpx=np.reshape(simX[is,:,:],(1,2,1))
        yt=lstm_model.predict(inpx)
        simX[is+1,1]=(yt+np.random.normal(loc=0,scale=std_err,size=1))
        simY[is+1]=simX[is+1,1]
        simX[is+1,0]=simX[is,1]
    simZ[:,it]=simY*(maxX-minX)+minX
```

12.1.5 Matlab Coding

Another comparable program is Matlab for deep learning. It is assumed that the deep learning toolbox is installed in the version of 2019. The Colorado River streamflow is uploaded at first as.

```
ColRiv=xlsread('Colorado_MonStream.csv')

ColRiv(1,:)=[];

Dtemp(:,:)=ColRiv*1233/10^9;% acre feet to 10^9 m^3
```

The first line of the uploaded data was removed, and its unit was converted from acre-feet to 10^9 m^3. The monthly data were transformed to the standard normal variable with the copula-based transformation (Lee and Salas 2011) as

$$\hat{y} = F_{norm}^{-1}(F_{gam}(x)) \tag{12.2}$$

where F_{gam} is the CDF of gamma distribution as in Eq. (2.34) and F_{norm} is the CDF of the standard normal distributions as in Eq. (2.30). This data transformation is done for each month as follows:

```
for im=1:12

  [aa]=gamfit(Dtemp(:,im))

  Stat.GamPar.alpha(im)=aa(1);

  Stat.GamPar.beta(im)=aa(2);

  Data_normv(:,im)=norminv(gamcdf(Dtemp(:,im),Stat.GamPar.alpha(im),Stat.GamPar.beta(im)));

end


data_stz(1:sz_org(1)*12)=Data_normv';
```

The dataset is split into training part and testing part with 60 and 40% each as.

```
numTimeStepsTrain = floor(0.6*numel(data_stz));
XTrain = data_stz(1:numTimeStepsTrain);
YTrain = data_stz(2:numTimeStepsTrain+1);
XTest = data_stz(numTimeStepsTrain+1:end-1);
YTest = data_stz(numTimeStepsTrain+2:end);
```

Note that the simulation model is set as

$$Y_t = f(y_{t-1}) + \varepsilon_t \tag{12.3}$$

Therefore, just one input and output variables are employed with four hidden units as.

```
inputSize = 1;
numResponses = 1;
numHiddenUnits = 4;
```

The model structure is built:

```
layers = [ ...
  sequenceInputLayer(inputSize)
  lstmLayer(numHiddenUnits)
  fullyConnectedLayer(numResponses)
  regressionLayer];
```

This indicates one input layer and LSTM layer, followed by the output layer. The option for training is also set as follows:

```
opts = trainingOptions('adam', ...
  'MaxEpochs',250, ...
  'GradientThreshold',1, ...
```

```
'InitialLearnRate',0.005, ...
'LearnRateSchedule','piecewise', ...
'LearnRateDropPeriod',125, ...
'LearnRateDropFactor',0.2, ...
'Verbose',0)%'Plots','training-progress');
```

Here, the number of epochs (“MaxEpochs”) is 250 and the learning rate (“InitialLearnRate”) in Eq. (5.2) is 0.005.

Subsequently, the model is fitted and the prediction values for the testing dataset are estimated as.

```
net = trainNetwork(XTrain,YTrain,layers,opts);
for iym = 1:length(XTest)
  [net,YPred1_cur(iym)] = predictAndUpdateState(net,XTest(iym));
end
  err_std=std(YPred1_cur-YTest);% Estimate the error variance
```

Its variance (err_std) is estimated by employing the estimated predicted value.

Eventually, the simulation is performed. The first value is selected from the training data and its state is updated accordingly as.

```
InitSt=unidrnd(length(XTrain)-1);
Yinit=XTtrain(InitSt*12+1);
[net] = predictAndUpdateState(net,Yinit);
Ysim_P= Yinit
```

Then, $y_{t-1} = \widehat{Y}_t$, that is, Ysim_P = Yinit. Then, simulation is done by updating the predicted value and adding the variance as $Y_t = f(y_{t-1}) + \varepsilon_t$.

```
[net,Ysim_pred(iym)] = predictAndUpdateState(net,Ysim_P);
Ysim(iym)=Ysim_pred(iym)+err_std*randn;
```

The simulated value is back-transformed for each month as

$$x = F_{gam}^{-1}(F_{norm}(x))$$

```
Ysim_org=gaminv(normcdf(Ysim),Stat.GamPar.alpha(monD1),Stat.GamPar.beta(monD1));
```

The total Matlab programming is as follows.

[Matlab 12.1] Stochastic Simulation of Colorado River

```
clear
ColRiv=xlsread('Colorado_MonStream.csv')
ColRiv(1,:)=[];
Dtemp(:,:)=ColRiv*1233/10^9;% acre feet to m^3
sz_org=size(Dtemp);
monD=repmat(1:12,sz_org(1),1);
monD1(1:sz_org(1)*12)=monD';%Month Index
Dtemp_yr=sum(Dtemp');
%% Basic Stat----------------------
for im=1:12
    [aa]=gamfit(Dtemp(:,im))
    Stat.GamPar.alpha(im)=aa(1);
    Stat.GamPar.beta(im)=aa(2);
      Data_normv(:,im)=norminv(gamcdf(Dtemp(:,im),Stat.GamPar.alpha(im),
      Stat.GamPar.beta(im)));
end
data_stz(1:sz_org(1)*12)=Data_normv';
```

```
%% Data Setup for LSTM Training
numTimeStepsTrain = floor(0.6*numel(data_stz));
XTrain = data_stz(1:numTimeStepsTrain);
YTrain = data_stz(2:numTimeStepsTrain+1);
XTest = data_stz(numTimeStepsTrain+1:end-1);
YTest = data_stz(numTimeStepsTrain+2:end);

inputSize = 1;
numResponses = 1;
numHiddenUnits = 4;
layers = [ ...
    sequenceInputLayer(inputSize)
    lstmLayer(numHiddenUnits)
    fullyConnectedLayer(numResponses)
    regressionLayer];
opts = trainingOptions('adam', ...
    'MaxEpochs',250, ...
    'GradientThreshold',1, ...
    'InitialLearnRate',0.005, ...
    'LearnRateSchedule','piecewise', ...
    'LearnRateDropPeriod',125, ...
    'LearnRateDropFactor',0.2, ...
    'Verbose',0)%'Plots','training-progress');
```

[Matlab 12.1] Continue

```
net = trainNetwork(XTrain,YTrain,layers,opts);
for iym = 1:length(XTest)
    [net,YPred1_cur(iym)] = predictAndUpdateState(net,XTest(iym));
end
    err_std=std(YPred1_cur-YTest);% Estimate the error variance

N.it=100;
Ysim_all(N.it,sz_org(1)*12)=0;
N.sim=sz_org(1)*12;%Simualtion Record Length

for it = 1:N.it
    InitSt=unidrnd(length(XTrain) -1);
    Yinit=Yall(InitSt);
    [net] = predictAndUpdateState(net,Yinit);
    Ysim_P=Yinit(end);
    for iym=1:N.sim
        [net,Ysim_pred(iym)] = predictAndUpdateState(net,Ysim_P);
        Ysim(iym)=Ysim_pred(iym)+err_std*randn;
        Ysim_P=Ysim(iym);
    end

      Ysim_org=gaminv(normcdf(Ysim),Stat.GamPar.alpha(monD1),Stat.GamPar
      .beta(monD1));
    Ysim_all(it,:)= Ysim_org;

end
```

12.2 Forecasting Daily Temperature with LSTM

In this chapter, LSTM assisted with the Genetic Algorithm (GA) to predict one-day-ahead and multi-day-ahead maximum temperature is presented. This is a challenging task when working with time-series data. A weather time-series dataset recorded at Cheongju station, South Korea, is used. The dataset consists of daily maximum temperature from the year 1976 to 2015 and is used to build a model that takes some data from the recent past observations as input and predicts the maximum temperature 1 day and 15 days in the future. The data format is in excel. For example, the data in each line has Year-Month-Day-Temperature (°C), "1976 11 17 0.5."

The procedure can be divided into three parts: to (1) preprocess the data, (2) estimate the parameter set including the hyperparameters, and (3) apply the estimated parameter set to predict the future. Here, in this temperature prediction, the estimation of the hyperparameters is focused.

12.2.1 *Preparing the Data*

The stationary condition of the target data is critical to model for time series forecasting and is likely to lead to more skillful forecasts. In another way, it must be expected that there is no long-term trend or cyclicity of the dataset. To satisfy these conditions, the temperature data is divided into four seasons, namely winter (from December to February), spring (from March to May), summer (from June to August), and autumn (from September to November). The GA-LSTM model is used to make temperature prediction for four seasons. However, through this example, only the results and code for summer are shown and one can use the codes to predict for the other seasons in a similar way.

The data for time series forecasting needs to be scaled at first when training a neural network, since large value inputs may slow down the network's learning and convergence. In some cases, it can prevent the network from learning the problem effectively. Therefore, in this example, the data is preprocessed by subtracting the mean of time series and dividing by the standard deviation. Also, the whole observed data must be split into three parts as the first 80% of data (1976–2007) as training data, and compute the mean and standard deviation only for this set. Meanwhile, the remaining 20% (2008–2015) is used as the test set. Furthermore, a validation set containing 20% of the training data (2001–2007) is held out to validate the performance of the model. After that, GA and LSTM are integrated to find the customized architecture of the LSTM model (the number of hidden neurons and epochs) for the forecasting of one-day-ahead maximum temperature.

12.2.2 *Methodology*

The prediction process consists of two phases as the selection of optimal hyperparameters and prediction with the selected hyperparameters. The first phase of the experiment involves selecting the optimal hyperparameters (i.e., number of hidden neurons and epochs) for the LSTM model using GA in the current example. The training set to fit models was used and prediction corresponding to the validation set was made, and then the root mean square error (RMSE) of the predicted values was measured. For one day prediction, the time series was split into input and output using the lags time technique. Particularly, past 7-day daily data values (x_{t-6}, ..., x_{t-1}, x_t) were employed to predict the daily maximum temperature for the next time (x_{t+1}). The tanh function limiting the inputs to a range $[-1; 1]$ was used as an activation function of the input and hidden neurons, as explained in Eq. (4.5).

Meanwhile, a linear function is defined as an activation function of output neurons in models. The initial weights of the network are set as random values, and a gradient-based "Adam" optimizer is applied to train the LSTM model. The GA solution would be decoded to get an integer number of epochs and a number of hidden neurons to train models. After that, the best value of the hyperparameters of the model was determined by selecting the smallest RMSE of the validation set (optimum model) using the Genetic Algorithm (GA). Genetic parameters, such as crossover rate, mutation rate, and population size, may affect the outcome in order to get the optimal solution to the problem. In this example, a population size of 10, 0.5 for crossover rate, and 0.1 for mutation rate were employed.

The number of neurons in each hidden layer varies from 1 to 20, and the number of epochs from 20 to 300. Additionally, the number of hidden layers from 1 to 3 was assessed. The number of generations is allocated as 10 as a termination condition. The GA is implemented by using the Distributed Evolutionary Algorithms in Python (DEAP) library. Finishing this phase, one can find the best architecture of the LSTM model corresponding to different cases of hidden layers.

In the second phase, the best-defined hyper-parameters corresponding to different hidden layers of LSTM was applied to forecast maximum temperature. In this stage, the LSTM model was fitted with optimal hyper-parameters for both training and validation sets, and then the test set was used for forecasting. Experiments were performed 1 day and 15 days in the future. At first, a one-day-ahead maximum temperature was forecasted using seven previous observations.

To estimate the prediction accuracy and evaluate the performance of the forecast, the root mean square error (RMSE) was calculated as

$$RMSE = \sqrt{\frac{1}{n}\sum_{t=1}^{n}(x_t - \hat{x}_t)^2} \tag{12.4}$$

where x_t is the current true value, $\hat{x}_t$ is its predicted value, and n is the total number of testing data. RMSE is commonly used to calculate the difference between predicted and observed values, and the performance of a model will be better when the value of RMSE is close to zero.

The optimal LSTM model corresponding to various hidden layers (i.e., 1–3 hidden layers) was trained to the data to select the best-hidden layer for one-day-ahead prediction. For example, in summer, after running the GA algorithm, the best number of hidden neurons (3) and epochs (25) for the LSTM model that had 1 hidden layer was found. These values were applied to train the data and the RMSE of the model on the test set was estimated. Similarly, the RMSE in the case of 2 and 3 hidden layers was estimated, and finally, the RMSEs were compared among them to obtain the best-hidden layer of the LSTM model that produced the smallest RMSE for the one-day-ahead prediction problem.

For prediction for the next 15 days, the iteration technique was employed. It means that when the first prediction (x_{t+1}) was obtained, this value was inserted into the input series to obtain the prediction x_{t+2} for the next time step. Finally, we had a sequence $[x_{t+1}, x_{t+2}, ..., x_{t+15}]$ that was predicted after 15 iteration steps. The model employed different numbers of lags (7–36) for each season to find the best lags based on the RMSE of 15-day-ahead prediction. The results from some trial runs showed that when the number of lags was increased beyond this level, the error of the model did not decrease significantly so that the experiment was stopped at 36 lags.

12.2.3 Results

The results of the LSTM model for 1 day and multi-days ahead of maximum temperature were plotted. It can be noticed from Fig. 12.5 that the divergence between prediction and observation values of maximum temperature increased for longer lead times. Furthermore, the figure depicted that most of the low values were overestimated, whereas high values were underestimated when predicting long days ahead.

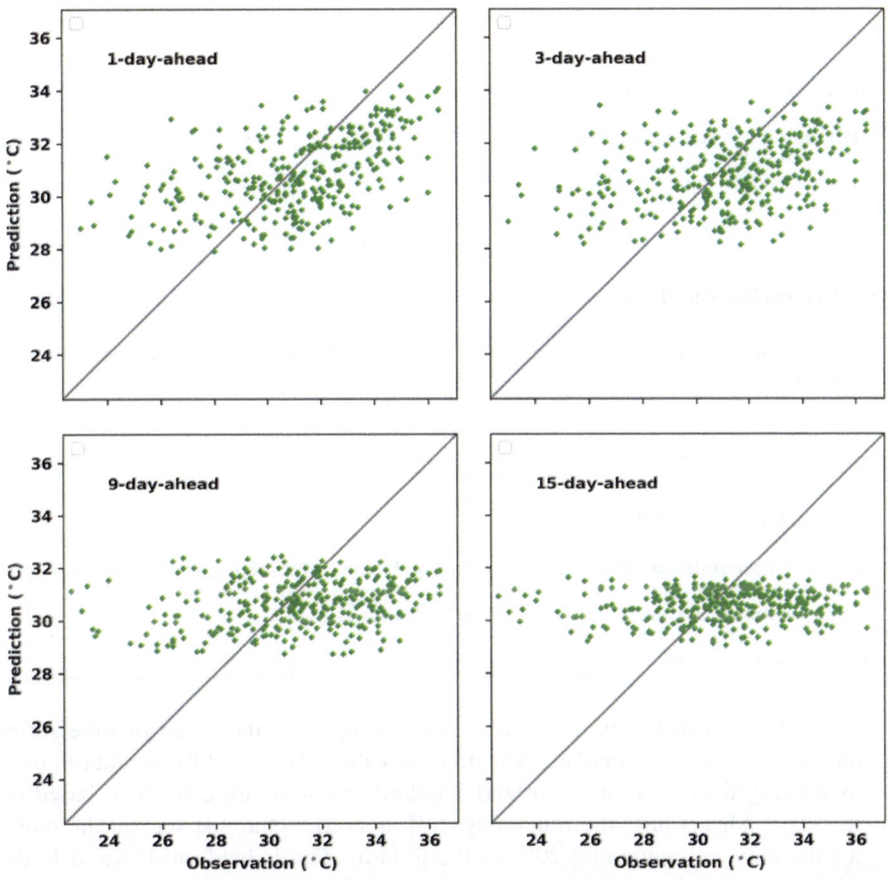

Fig. 12.5 Scatter plot of prediction versus observation for multi-days ahead using LSTM for the test set in summer

12.2.4 Python Coding

The LSTM is implemented using Keras and the GA is implemented by using Distributed Evolutionary Algorithms in Python (DEAP) library.

(a) **Importing the necessary libraries**

```
import random
import numpy as np
import pandas as pd
from deap import base, creator, tools, algorithms
from sklearn.metrics import mean_squared_error
from keras.models import Sequential
from keras.layers import Dense,LSTM,Activation
from keras.optimizers import adam
from math import sqrt
np.random.seed(42)
```

(b) **Preparing the data**

– Upload data using the function *read_excel()* in Pandas and concatenate data for summer.

```
df=pd.read_excel('cheongju.xlsx', header=None)
df.columns = ['year', 'month', 'day', 'Tmax']
summer=pd.concat([df[df.month==6], df[df.month==7], df[df.month==8]], ignore_index=True)
summer= summer.sort_values(by=['year', 'month', 'day'])
dataset=summer['Tmax'].values
```

– Next, the data would be preprocessed by subtracting the mean of time series and dividing by the standard deviation. Notably, the first 80% of data is used as training data, and the mean and standard deviation are calculated based on this data. Meanwhile, the remaining 20% is used as the test set. Furthermore, a validation set containing 20% of the training data is held out to validate the performance of the model.

```
num_train=int(0.8*len(dataset))

mean = dataset[:num_train].mean(axis=0)

dataset -= mean

std = dataset[:num_train].std(axis=0)

dataset /= std

train=dataset[0:num_train]

num_trains=int(0.8*len(train))

#validation set

val=train[num_trains:num_train]

#test set

test=dataset[num_train:]

#training set

train2=train[:num_trains]
```

– After that, a function to prepare the dataset is defined. The input would include the number of past observations (i.e., lags) and the output would be the maximum temperature in the next day.

```
def create_dataset(dataset, lags):

  dataX, dataY = [], []

  for i in range(len(dataset)-lags-1):

          a = dataset[i:(i+lags)]

          dataX.append(a)

          dataY.append(dataset[i + lags])

  return np.array(dataX), np.array(dataY)
```

(c) **Building model**

An LSTM model with one hidden layer is defined by function *fit_lstm ()* to forecast one-day-ahead maximum temperature using seven past observations.

```
def fit_lstm (X,y, batch_size, epochs, neurons):

    model = Sequential()

    model.add(LSTM(neurons[0],input_shape=( X.shape[1], X.shape[2])))

    model.add(Dense(1))

    model.add(Activation('linear'))

    model.compile(loss='mean_squared_error', optimizer='adam')

    history=model.fit(X, y, epochs=epochs,batch_size=batch_size ,verbose=0,shuffle=False)

    return model
```

(d) **Building fitness function**

The *train_evaluate()* function created the LSTM model for a given individual and returns its RMSE value for the validation set. The best value of hyperparameters of the model was determined by selecting the smallest RMSE on the validation set (optimum model) using the GA.

```
def train_evaluate(ga_individual_solution):

   #Decode genetic algorithm solution to integer for number of epoch and number of hidden
neurons

    n_epoch = ga_individual_solution[0]

    neuron1 = ga_individual_solution[1]


    print('\nnum of epochs = ',n_epoch,", neurons = [",neuron1,"]")

    neurons = []

    neurons.append(neuron1)


    #Train LSTM model on training set and prediction on validation set based on new epochs

    #and hidden neurons

    X, y = create_dataset(train2, lags=7)

    X = X.reshape(X.shape[0], X.shape[1],1 )

    lstm_model = fit_lstm(X,y, 100,n_epoch, neurons)
```

```
    X_val, y_val = create_dataset(val, lags=7)

    X_val = X_val.reshape(X_val.shape[0], X_val.shape[1],1 )

    y_pred=lstm_model.predict(X_val)


    #Calculate the RMSE score as fitness score for GA

    y_prescaled=y_pred*std+mean

    y_true=y_val*std+mean

    rmse_val = sqrt(mean_squared_error(y_true, y_prescaled))


    print('test RMSE: %.4f' % rmse_val)


    return rmse_val,
```

(e) **Set up GA using DEAP**

Genetic parameters, such as crossover rate, mutation rate, and population size, may affect the outcome in order to get the optimal solution to the problem. In this example, a population size of 10, 0.5 for crossover rate, and 0.1 for mutation rate were used. The number of neurons in each hidden layer varies from 1 to 20, and the number of epochs from 20 to 300. The number of generations is allocated as 10 as a termination condition.

```
# number of solutions in the population

population_size = 10

# number of time we generate new population

num_generations = 10


# create a minimizing fitness function because we want to minimize RMSE

creator.create('FitnessMax', base.Fitness, weights = (-1.0,))

creator.create('Individual', list , fitness = creator.FitnessMax)
```

```
# create an object of Toolbox class

toolbox = base.Toolbox()


toolbox.register("attr_int1", random.randint, 20,300)

toolbox.register("attr_int2", random.randint,1,20)
toolbox.register("individual", tools.initCycle, creator.Individual,

        ( toolbox.attr_int1,toolbox.attr_int2), n=1)


toolbox.register('population', tools.initRepeat, list , toolbox.individual)

toolbox.register('mate', tools.cxTwoPoint)

toolbox.register('mutate', tools.mutUniformInt,low = [20,1],up = [300,20],indpb = 0.6)

toolbox.register('select', tools.selRoulette)

toolbox.register('evaluate', train_evaluate)


# create population by calling population function

population = toolbox.population(n = population_size)


# start GA

r = algorithms.eaSimple(population, toolbox, cxpb = 0.5,

            mutpb = 0.1, ngen = num_generations, verbose = False)
```

These codes will create a minimizing fitness function (i.e., minimize RMSE) by building a *FitnessMax* class with the value of *weights* to be the tuple (−1.0) and the "Individual" will be presented as an array.

In this block, a DEAP container called *Toolbox* will be created to define the Individual and create the population. The first attribute "attr_int1" (i.e., the number of epochs) in the individual will be produced as a random integer between 20 and 300. The second attribute, "*attr_int2* (i.e., the number of hidden neurons) in the individual, will be generated between 1 and 20. Mating will use a two-point crossover. An individual will be mutated by replacing attributes, with probability $indpb = 0.6$, by an integer uniformly drawn from *low* to *up* inclusively and use the roulette wheel selection for selecting the parents.

(f) **Finding the best LSTM model**

The best architecture of the LSTM model (i.e., number of epochs and hidden neurons) corresponding to one hidden layer was found as follows:

```
best_individuals = tools.selBest(population,k = 1)

best_epochs = None

best_neuron1 = None


print("\nBest solution is:")

for bi in best_individuals:

  best_epochs = bi[0]

  best_neuron1 = bi[1]

  print('\n epochs = ',best_epochs,", neurons = [",best_neuron1,"]")
```

(g) **Forecasting one-day-ahead maximum temperature**

The best LSTM solution would be implemented to forecast the maximum temperature for the next day on the test set. It is noted that in this stage, the LSTM model with optimal hyper-parameters should be retrained with both training and validation sets, and then should use the test set for forecasting.

```
#Apply the best hyper-parameter to train LSTM model

#the best number of hidden neurons

neurons= best_neuron1

#the best number of epochs

epochs= best_epochs

#retrain LSTM with training and validation set
X, y = create_dataset(train, lags=7)

model = lstm_model(X, y, 100, epochs, neurons)

#prediciton 1 day ahead

X_test, y_test = create_dataset(test, lags=7)

X_test1 = X_test.reshape(X_test.shape[0], X_test.shape[1],1 )

y_pred=model.predict(X_test1)

y_prescaled=y_pred*std+mean

y_true=y_test*std+mean
```

```
rmse_test = sqrt(mean_squared_error(y_true, y_prescaled))

print('Test RMSE: %.3f' % rmse_test)
```

(h) **Forecasting multi-day-ahead maximum temperature**

- It is noted that the LSTM model will be retrained with both training and validation sets again when applying the best number of lags.

```
#Applying the best number of lags to train LSTM model

best_lags=35

X, y = create_dataset(train, best_lags)

X = X.reshape(X.shape[0], X.shape[1],1 )

model = fit_lstm(X, y, 100, best_epochs, [best_neuron1])
```

- Next, define a function to split the time series into samples based on the number of new lags and number of outputs.

```
def split_sequences(data, n_steps_in, n_steps_out):
  X, y = list(), list()

  for i in range(len(data)):

          # find the end of this pattern

          end_ix = i + n_steps_in

          out_end_ix = end_ix + n_steps_out

          # check if we are beyond the dataset

          if out_end_ix > len(data)-1:

                  break

          # gather input and output parts of the pattern

          seq_x, seq_y = data[i:end_ix], data[end_ix:out_end_ix]

          X.append(seq_x)

          y.append(seq_y)

  return np.array(X), np.array(y)
```

– A function *moving_window_preds()* is created to forecast the future values of maximum temperature. The *n_pred_ahead* represents the number of future predictions and this coincides with the number of windows that we will move forward on the test data.

```
def moving_window_preds(input_data,n_pred_ahead):

  # Creating the first test window

  moving_window = input_data

  pred_moving = np.zeros((len(input_data),1))


  for i in range(n_pred_ahead):

    pred_one_step = model.predict(moving_window)

    pred_moving=np.concatenate((pred_moving,pred_one_step),axis=1)

    # Reshaping the prediction to 3D array
    pred_one_step = pred_one_step.reshape(len(pred_one_step),1,1)

    # This is the new moving test window, where the first element from the window has been
removed

    #and the prediction has been appended to the end

    moving_window = np.concatenate((moving_window[:,1:,:], pred_one_step), axis=1)


  pred_moving =  pred_moving[:,1:]*std+mean


  return pred_moving
```

Finally, the prediction from 1 to 15 days in advance of maximum temperature were implemented, based on the optimum LSTM as follows:

```
error=[]

for i in range(1, 16):

  lags, output = best_lags, i

  X, y = split_sequences(test,lags, output)

  X = X.reshape(X.shape[0], X.shape[1],1 )

  preds_moving = moving_window_preds(X, i)
```

```
x_fut=preds_moving

y_fut=y*std+mean

rmse_test = sqrt(mean_squared_error(y_fut, x_fut))

error.append(rmse_test)

print('Test RMSE: %.3f' % error[i-1])
```

(i) **Visualization**

Let's plot the results of the LSTM model for 1 day and multi-days ahead of maximum temperature.

It can be noticed from Fig. 12.5 that the divergence between prediction and observation values of maximum temperature increased at the longer lead times. Furthermore, the figure depicted that most of the low values were overestimated, whereas high values were underestimated when predicting a long day ahead.

(j) **Python whole code**

The complete code example is listed below.

```
import random
import numpy as np
import pandas as pd
from deap import base, creator, tools, algorithms
from sklearn.metrics import mean_squared_error
from keras.models import Sequential
from keras.layers import Dense,LSTM,Activation
from keras.optimizers import adam
from math import sqrt
np.random.seed(42)

df=pd.read_excel(r'E:\KIỀU
TRANG\MASTER_KOREA\PYTHON\python_file\Korea_LSTM\Cheongju_model\cheongju.xlsx',
header=None)
df.columns = ['year', 'month', 'day', 'Tmax']
summer=pd.concat([df[df.month==6], df[df.month==7], df[df.month==8]], ignore_index=True)
summer= summer.sort_values(by=['year', 'month', 'day'])
dataset=summer['Tmax'].values
num_train=int(0.8*len(dataset))
mean = dataset[:num_train].mean(axis=0)
dataset -= mean
std = dataset[:num_train].std(axis=0)
dataset /= std
train=dataset[0:num_train]
num_trains=int(0.8*len(train))
#validation set
val=train[num_trains:num_train]
#test set
```

```
test=dataset[num_train:]
#training set
train2=train[:num_trains]

def create_dataset(dataset, lags):
        dataX, dataY = [], []
        for i in range(len(dataset)-lags-1):
                a = dataset[i:(i+lags)]
                dataX.append(a)
                dataY.append(dataset[i + lags])
        return np.array(dataX), np.array(dataY)
```

```
def fit_lstm (X,y, batch_size, epochs, neurons):
   model = Sequential()
   model.add(LSTM(neurons[0],input_shape=( X.shape[1], X.shape[2])))
   model.add(Dense(1))
   model.add(Activation('linear'))
   model.compile(loss='mean_squared_error', optimizer='adam')
   history=model.fit(X, y, epochs=epochs,batch_size=batch_size ,verbose=0,shuffle=False)
   return model

def train_evaluate(ga_individual_solution):

   #Decode genetic algorithm solution to integer for number of epoch and number of hidden
neurons
   n_epoch = ga_individual_solution[0]
   neuron1 = ga_individual_solution[1]

   print('\nnum of epochs = ',n_epoch,", neurons = [",neuron1,"]")

   neurons = []
   neurons.append(neuron1)

#Train LSTM model on training set and prediction on validation set based on new epochs and
hidden neurons
   X, y = create_dataset(train2, lags=7)
   X = X.reshape(X.shape[0], X.shape[1],1 )
   lstm_model = fit_lstm(X,y, 100,n_epoch, neurons)

   X_val, y_val = create_dataset(val, lags=7)
   X_val = X_val.reshape(X_val.shape[0], X_val.shape[1],1 )
   y_pred=lstm_model.predict(X_val)

   #Calculate the RMSE score as fitness score for GA
   y_prescaled=y_pred*std+mean
   y_true=y_val*std+mean
   rmse_val = sqrt(mean_squared_error(y_true, y_prescaled))

   print('test RMSE: %.4f' % rmse_val)

   return rmse_val,
```

```
# num of solutions in the population
population_size = 10
# num of time we generate new population
num_generations = 10

# create a minimizing fitness function, cause we want to minimize RMSE
creator.create('FitnessMax', base.Fitness, weights = (-1.0,))
creator.create('Individual', list , fitness = creator.FitnessMax)

# create an object of Toolbox class
toolbox = base.Toolbox()

toolbox.register("attr_int1", random.randint, 20,300)
toolbox.register("attr_int2", random.randint,1,20)
toolbox.register("individual", tools.initCycle, creator.Individual,
        ( toolbox.attr_int1,toolbox.attr_int2), n=1)

toolbox.register('population', tools.initRepeat, list , toolbox.individual)
toolbox.register('mate', tools.cxTwoPoint)
toolbox.register('mutate', tools.mutUniformInt,low = [20,1],up = [300,20],indpb = 0.6)
toolbox.register('select', tools.selRoulette)
toolbox.register('evaluate', train_evaluate)

# create population by calling population function
population = toolbox.population(n = population_size)

# start GA
r = algorithms.eaSimple(population, toolbox, cxpb = 0.5,
                mutpb = 0.1, ngen = num_generations, verbose = False)

best_individuals = tools.selBest(population,k = 1)
best_epochs = None
best_neuron1 = None

print("\nBest solution is:")
for bi in best_individuals:
        best_epochs = bi[0]
        best_neuron1 = bi[1]
        print('\n epochs = ',best_epochs,", neurons = [",best_neuron1,"]")
```

```
#Apply the best hyper-parameter to train LSTM model
#the best number of hidden neurons
neurons= best_neuron1
#the best number of epochs
epochs= best_epochs
#retrain LSTM with training and validation set
X, y = create_dataset(train, lags=7)
model = lstm_model(X, y, 100, epochs, neurons)

#prediciton 1 day ahead
X_test, y_test = create_dataset(test, lags=7)
X_test1 = X_test.reshape(X_test.shape[0], X_test.shape[1],1 )
y_pred=model.predict(X_test1)
y_prescaled=y_pred*std+mean
y_true=y_test*std+mean

rmse_test = sqrt(mean_squared_error(y_true, y_prescaled))
print('Test RMSE: %.3f' % rmse_test)
```

```
def split_sequences(data, n_steps_in, n_steps_out):
        X, y = list(), list()
        for i in range(len(data)):
                # find the end of this pattern
                end_ix = i + n_steps_in
                out_end_ix = end_ix + n_steps_out
                # check if we are beyond the dataset
                if out_end_ix > len(data)-1:
                        break
                # gather input and output parts of the pattern
                seq_x, seq_y = data[i:end_ix], data[end_ix:out_end_ix]
                X.append(seq_x)
                y.append(seq_y)
        return np.array(X), np.array(y)
```

```
def moving_window_preds(input_data,n_pred_ahead):
    # Creating the first test window
    moving_window = input_data
    pred_moving = np.zeros((len(input_data),1))

    for i in range(n_pred_ahead):
        pred_one_step = model.predict(moving_window)
        pred_moving=np.concatenate((pred_moving,pred_one_step),axis=1)
        # Reshaping the prediction to 3D array
        pred_one_step = pred_one_step.reshape(len(pred_one_step),1,1)
        # This is the new moving test window, where the first element from the window has been
removed
        #and the prediction has been appended to the end
        moving_window = np.concatenate((moving_window[:,1:,:], pred_one_step), axis=1)

    pred_moving =  pred_moving[:,1:]*std+mean

    return pred_moving

best_lags=35
X, y = create_dataset(train, best_lags)
X = X.reshape(X.shape[0], X.shape[1],1 )
model = fit_lstm(X, y, 100, best_epochs, [best_neuron1])

#predicition 15 days ahead
error=[]
for i in range(1, 16):
    lags, output = best_lags, i
    X, y = split_sequences(test,lags, output)
    X = X.reshape(X.shape[0], X.shape[1],1 )
    preds_moving = moving_window_preds(X,i)
    x_fut=preds_moving
    y_fut=y*std+mean
    rmse_test = sqrt(mean_squared_error(y_fut, x_fut))
    error.append(rmse_test)
    print('Test RMSE: %.3f' % error[i-1])
```

12.3 Exercises

1. Program a recurrent neural network (RNN) model to fit the Colorado River monthly streamflow data with Python and Matlab. Employ the previous three-month data as predictors $\left(\mathbf{X}_t = \left[y_{t-1}, y_{t-2}, y_{t-3}\right], \mathbf{Y}_t = y_t\right)$
   (a) Estimate the error variance of the RNN model with the errors of testing data
   (b) Simulate the Colorado River data for the same record length of historical data and 100 series
   (c) Compare the estimated statistics of mean, standard deviation, skewness, maximum, and minimum at each month (see Fig. 12.2 for detail).

2. Build a prediction model of LSTM for daily maximum temperature with Python and Matlab. Employ the previous five days data as predictors $\left(\mathbf{X}_t = \left[y_{t-1}, y_{t-2}, y_{t-3}, y_{t-4}, y_{t-5}\right], \mathbf{Y}_t = y_t\right)$ and the same dataset of South Korea in the case study of Sect. 12.2.
   (a) Predict five previous days sequentially and by output data $\mathrm{Y}_t = \left[y_{t-1}, y_{t+1}, y_{t+2}, y_{t+3}, y_{t+4}, y_{t+5}\right]$
   (b) Plot the test data and observations (scatterplot) and estimate the coefficient of determination R^2.

References

Lee T, Salas JD (2006) Record extension of monthly flows for the Colorado River system. US Breau of Reclamation, Denver

Lee T, Salas JD (2011) Copula-based stochastic simulation of hydrological data applied to Nile River flows. Hydrol Res 42(4):318–330. https://doi.org/10.2166/nh.2011.085

Lee T, Salas JD, Prairie J (2010) An enhanced nonparametric streamflow disaggregation model with genetic algorithm. Water Resour Res 46(8). https://doi.org/10.1029/2009WR007761

Lee T, Shin JY, Kim JS, Singh VP (2020) Stochastic simulation on reproducing long-term memory of hydroclimatological variables using deep learning model. J Hydrol 582. https://doi.org/10.1016/j.jhydrol.2019.124540

Salas JD (1993) Analysis and modeling of hydrologic time series. In: Maidment DR (ed) Handbook of hydrology. McGraw-Hill, New York, pp 19.1–19.72

Chapter 13
Environmental Applications of Deep Learning

Abstract A case study of CNN, a spatial deep learning model, is presented in this chapter for analyzing water quality with remote sensing data. Airborne remote sensing of cyanobacteria with multispectral/hyperspectral sensors is employed for input data and its water quality as output data.

13.1 Remote Sensing of Water Quality Using CNN

13.1.1 Introduction

Freshwater cyanobacterial bloom is one of the major water quality issues because it substantially deteriorates water quality and public health (Clark et al. 2017). Algal bloom, mainly caused by a combination of non-point or/and point sources, has been regarded as a global water quality issue. Currently, longer water retention times due to the construction of weirs accelerate water quality degradation and intensive algal or cyanobacterial blooms in South Korea (Yoon et al. 2015).

Here, airborne remote sensing of cyanobacteria with multispectral/hyperspectral sensors has been suggested as a promising technology due to its ability to effectively monitor a broader area of water resources with higher optical/spatial resolution (Glasgow et al. 2004; Lee et al. 2001). Multispectral or hyperspectral remote sensing has demonstrated its ability to monitor water quality in natural surface waters. In particular, airborne remote sensing with these sensors has successfully monitored the spatiotemporal water quality in many different waterbodies (Pyo et al. 2019).

With advances in optical monitoring on cyanobacteria, many researchers have proposed empirical or semi-analytical algorithms to retrieve cyanobacterial pigments for many different waterbodies. C*hlorophyll-a* (Chl-a, the pigment of phytoplankton biomass) and *phycocyanin* (PC, unique pigment of cyanobacteria) are target pigments to be quantified (Duan et al. 2012). These algorithms have been widely used to quantify Chl-a and PC in many different surface waters. However, they are highly dependent on the optical complexity of targeting water (i.e., concentration and composition of phytoplankton, suspended solids, and colored dissolved organic matter).

T. Lee et al., *Deep Learning for Hydrometeorology and Environmental Science*, Water Science and Technology Library 99,
https://doi.org/10.1007/978-3-030-64777-3_13

Atmospheric interference and its influence on algal retrieval have been issued for a long time and many researchers have reported their high complexities (Kimes et al. 2000). As well, airborne monitoring with a hyperspectral sensor provides a substantial amount of optical information about surface waters, but selecting a few optimal wavelengths is always problematic because it loses too much information.

Due to these imitations of existing algorithms, the deep learning technique could be an alternative way to quantify cyanobacterial pigments because it has demonstrated satisfactory performance for dealing with multi-dimensional images in various research fields (Huang et al. 2018). Among deep learning techniques, convolutional neural networks (CNNs) are regarded as an adequate method in the handling of these images, especially for resolving multi-classification problems (Kellenberger et al. 2018). The CNN model extracts significant or desired features from multiple imageries throughout sequential operations, such as convolutional layers, pooling layers, etc.

Eventually, this leads to achieving the effective pretreatment of multi-dimensional images without much loss of spatial and temporal information. Currently, it has successfully proved its ability to extract or classify features from hyperspectral imageries in many different fields. In particular, Kumar and Bhandarkar (2017) applied the CNN model to automatically recognize harmful cyanobacteria bloom from smartphone image textures. This study implies that CNN might be a proper technology to monitor algal blooms in surface waters, incorporated with optical sensing observation. Previous applications of CNN in handling hyperspectral imageries have mostly focused on classification rather than regression of target variables (Hu et al. 2015). In particular, it has rarely been implemented to estimate cyanobacteria from hyperspectral images.

A patch-wise approach for input composition in CNN-based regression was introduced to achieve computational effectiveness (Chen et al. 2016). The size of the input patch plays an important role for extracting the significant feature using CNN because of high complexity and heterogeneity from multidimensional image pixels (Zhang et al. 2018). Therefore, selecting an adequate size of the patch is the most critical step to extract high-level features and it significantly influences the spatial distribution of cyanobacteria. Deeper studies are required to investigate the effect of input patch size on CNN regression performance and to determine the optical size for the superior performance. Here, this application aims to (1) make an example on how a CNN model can be generated with hyperspectral imageries; (2) to train and validate the CNN model to quantify the concentrations of PC and Chl-a throughout mapping images; and (3) to compare the accuracies of the CNN model with conventional optical models.

13.1.2 Study Area and Monitoring

The study area is Baekje weir in Geum River (36°31′87.75′N 126°93′ 90.52″E) as shown in Fig. 13.1 with 23 km of stream length, approximately 8,000 km^2 of basin

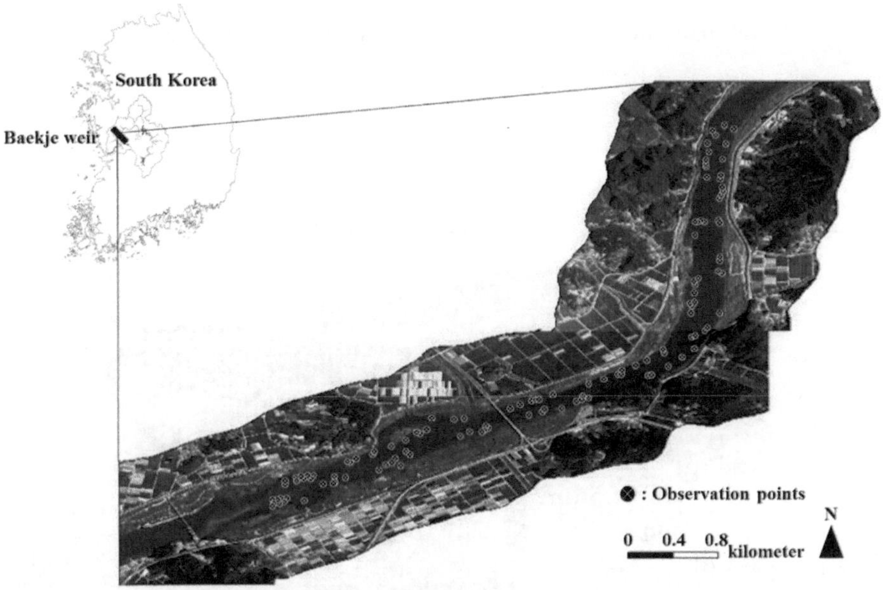

Fig. 13.1 Study area: Baekje weir with sampling point of each monitoring period. Red box indicates the location of the weir

area, and 12.2 million cubic meter of water storage. The weir has four different types of gates (fixed weir, moveable weir, small hydropower gate, and fishway). The major purposes of this weir are flooding control and water supply for domestic, industrial, and agricultural areas.

Since the weir construction in 2008, the algal/cyanobacteria has been a severe issue due to longer retention time. Furthermore, eutrophication is another trigger on the cyanobacteria blooms in the weir (Ministry of Environment, 2015). For instance, the weir has shown the above standard eutrophication index from January 2012 to August 2014 (Lee et al. 2014). In 2014, an intense algal/cyanobacteria bloom occurred because of the meteorological conditions as decrease of rainfall and increase of temperature, and solar intensity (Yoon et al. 2015).

13.1.3 Field Data Collection

From 2016 to 2017, there were a total of 9 events of on-site field experiments. Water samples from the field were collected and transferred into a laboratory for further analysis. Concurrently, an aircraft with the hyperspectral imagery sensor monitored the waterbody of the weir with 2-m spatial resolution, as shown in Fig. 13.2. Water surface reflectance data from the weir were collected with a wavelength range from 325 to 1075 nm taken from the aircraft. A FieldSpec HandHeld 2 Spectroradiometer

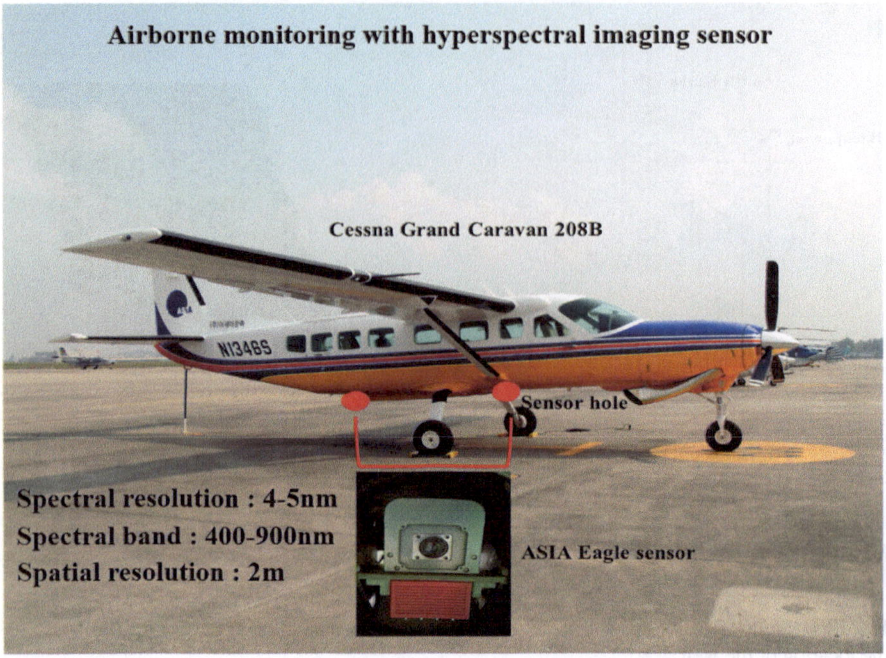

Fig. 13.2 Airborne monitoring with a hyperspectral sensor (ASIA Aero Survey Co.)

(ASD Inc., US) was used to measure the water surface reflectance, which is calculated from downwelling irradiance (Ed), downwelling radiance from the sky (Lsky), and upwelling radiance from the water surface (Lw).

The remote sensing reflectance (Rrs) was used to estimate the remote sensing reflectance in this study as:

$$Rrs = \frac{Lw - 0.025Lsky}{Ed} \tag{13.1}$$

The proposed azimuth and zenith angles by Mobley (1999) were adopted for this study. For every sampling point, two liters of water were taken for Chl-a and PC quantification. The solvent-extraction method using acetone and methanol was selected to measure Chl-a due to its fat-soluble feature. The freezing and thawing method was used to extract the PC due to its water-soluble property.

During in-situ monitoring, airborne monitoring was also conducted by ASIA Aero Survey Co. (Fig. 13.2). The aircraft with the hyperspectral sensor, AISA Eagle (SPECIM Inc., Finland) scanned the waterbody of the weir for a total of nine times in 2017 and 2018. The sensor has a spectral range from 400 to 970 nm with 4–5 nm spectral resolution. The geometric correction was performed and MODTRAN 6 was used to estimate the atmospheric correction parameters, eventually calculating the surface reflectance of images.

MODTRAN was developed by Spectral Science, Inc. and the Air Force Research Laboratory (AFRL). The MODTRAN code solves the radiative transfer function to generate physical parameters related to atmospheric correction, such as transmittance and spherical albedo. MODTRAN version 6 has a graphical user interface (GUI), making this software user-friendly. The MODTRAN 6 inputs were built to generate atmospheric correction parameters for the Baekje weir. The statistical band model and discrete ordinate radiative transfer were selected as the radiative transfer and multiple scattering algorithms, respectively. A mid-latitude summer atmospheric profile was used, and the CO_2 concentration was set to 400 ppmv. The rural boundary layer aerosol option was selected.

The geometry was specified by selecting the solar zenith angle and solar azimuth angle according to the date the field data were collected. This geometry was then used to set the solar geometry specification parameters for solar scattering. The zenith angle of the hyperspectral sensor was set to 180° since the sensor was installed on an aircraft perpendicular to the ground. Information on monitoring time and geographic coordinates (i.e., latitude and longitude) for each sampling point were written in the geometry input section. Lambertian spectral reflectance was defined in the surface specification option.

The spectral specification option sets the spectral range of the MODTRAN output from 400 to 800 nm. The MODTRAN 6 output provides atmospheric correction parameters, such as atmospheric path radiance, total solar flux, direct transmittance, diffuse transmittance, and spherical albedo. Based on simulated atmospheric correction parameters, the radiometric calibration converting the digital number (DN) of the raw hyperspectral image to at-sensor radiance was done following Duan et al. (2013). ρ_{surf} was calculated by inverting the radiometric calibration equation:

$$\rho_{\text{surf}} = \frac{\pi\left(L_{at-sensor} - L_{path}\right)}{\pi\left(L_{at-sensor} - L_{path}\right)S + T \times F_T} \tag{13.2}$$

where ρ_{surf} is the surface reflectance (sr^{-1}), $L_{at-sensor}$ is the total radiance through the sensor ($Wcm^{-2}sr^{-1}\ cm^{-1}$), L_{path} is the atmospheric path radiance ($WcFm^{-2}sr^{-1}\ cm^{-1}$), S is the spherical albedo of the atmosphere, T is the atmospheric transmittance, and F_T is the total solar flux at the ground ($Wcm^{-2}\ cm^{-1}$). The average value of calibrated radiative parameters was distributed on the hyperspectral image to calculate ρ_{surf} pixel-by-pixel.

13.1.4 Point-Centered Regression CNN (PRCNN)

A CNN was adopted to extract significant features to quantify PC and Chl-a pigments in the weir as shown in Fig. 13.3 with the CNN structure using hyperspectral images (i.e., digital number) and atmospheric parameters (i.e., total flux, diffuse transmittance, direct transmittance, spherical albedo, path radiance) estimated from

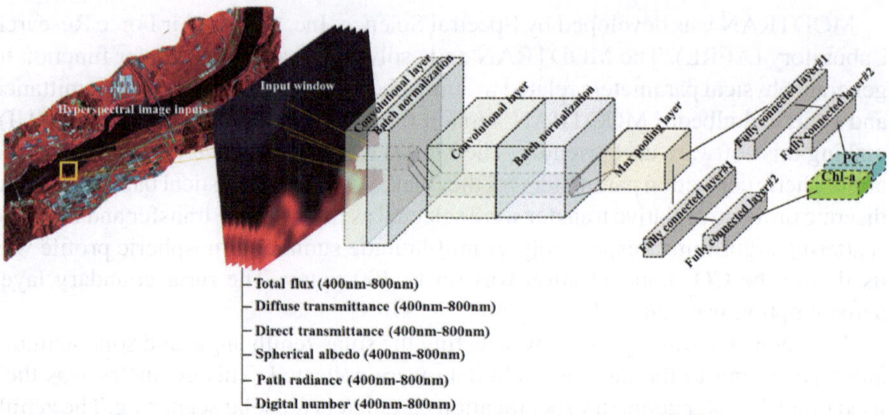

Fig. 13.3 Convolutional neural network architecture, input layer composed of atmospheric correction parameters and digital number, two convolutional layers, and two fully connected layers for phytoplankton pigment estimation

MODTRAN code. Here, each of 6 parameters had a total of 86 bands (from 400 to 800 nm). The input windows were adopted to generate smaller sectors from the stacked input images, containing monitoring points in the center of the sector. It refers to implementing a point-centered regression CNN (PRCNN) with the stacked input images. Thereby, the input size of input data is 119 × IW × IH × 516, where 119 is the number of sampling points and 516 is the total spectral depth. IW and IH represent the width and height of the extracted input image, respectively, ranging among 8 × 8, 64 × 64, and 128 × 128.

The input-window size in the convolutional process characterizes the nonlinear feature of the pigment distribution and the linear feature of the concentration. With those input images, two convolutional layers were applied and each convolutional layer sequentially connected to batch normalization, a max-pooling layer, and a dropout layer. The input images with filter size *FW* (the filter width) × *FH* (the filter height) and *N* (the filter depth) are first transferred to convolutional layers. According to the sizes of stride (*S*) and padding (*P*), the output size from the first convolutional layer changed to $((IW + 2P\text{-}FW)\text{-}1)/S \times ((IH + 2P\text{-}FH)\text{-}1)/S \times N$. After the convolutional layer, the batch normalization is applied to avoid early saturation of the activation function (e.g., vanishing-gradient issue) in the network and to achieve effective training (Chang and Chen 2015; Chen et al. 2016; Ioffe and Szegedy 2015). Then, the max-pooling layer is used to reduce the dimension of signals and to extract important features (Kalchbrenner et al. 2014) and the dropout layer is used to prevent overfitting of the network by randomly ignoring the nodes in the layer; thereby weight and bias in the nodes are not estimated in the backward network (Srivastava et al. 2014). After all, the processes mentioned above, the output signal is transferred to the two fully connected layers to produce PC and Chl-a concentration. An optimizer estimates the loss function (i.e., mean square error, MSE) in an attempt to minimize the discrepancy between simulated algal pigments and observations. Before this optimization, datasets are randomly separated into 70% training and 30% validation datasets.

13.1.5 Results and Discussion

The PRCNN model has input, including atmospheric parameters and optical information. MODTRAN 6 software produces atmospheric parameters, including total solar flux, direct transmittance, diffuse transmittance, path radiance, and spherical albedo (Pyo et al. 2018); the radiative transfer equation is used. Additionally, the digital numbers of the hyperspectral image are included to represent the optical properties of the waterbody. It is necessary to include two different types of information; light intensity (e.g., surface reflectance) and atmosphere condition (e.g., atmospheric interference). Each of six different inputs has 86 spectral bands from 400 to 800 nm. Here, it was assumed that inputs of the PRCNN model included the spatiotemporal variation of cyanobacteria pigments and atmospheric radiative properties; thereby, the model quantifies the spatiotemporal variation of pigments. In particular, the convolutional process with the filter could include the spatial properties of the optical information surrounding the monitoring point.

Additionally, uncertainty caused by sampling errors would be considered in the estimation of algal pigment by the convolutional process. Here, the filter depths composite atmospheric and optical information and use to quantify algal pigments. If the size of the input window is changed, the feature extracted by the CNN could be altered; the spatial and spectral features, which are used to quantify the pigments, would be enlarged or reduced. Here, we expect that the size of the input window substantially influences the performance of the CNN model.

The CNN model was trained to produce the spatiotemporal variation of cyanobacteria pigments using atmospheric radiative properties and hyperspectral images. As well, the convolutional process with different sizes of filters could mitigate the uncertainty of measurement generated from monitoring procedures. For example, the sampling boat kept moving due to the water flow during sampling events, and it might result in measurement uncertainty. Here, selecting the optimal filter size is a substantially influential process. The larger the size, the broader the features extracted from images. The filter depth compositing multiple input layers is also a significant parameter to be an effective training process. Consequently, the main purpose of the convolutional step in this study is to calculate the representative signal from atmospheric parameters and hyperspectral images to quantify algal pigments at the center pixel with consideration of measurement uncertainty.

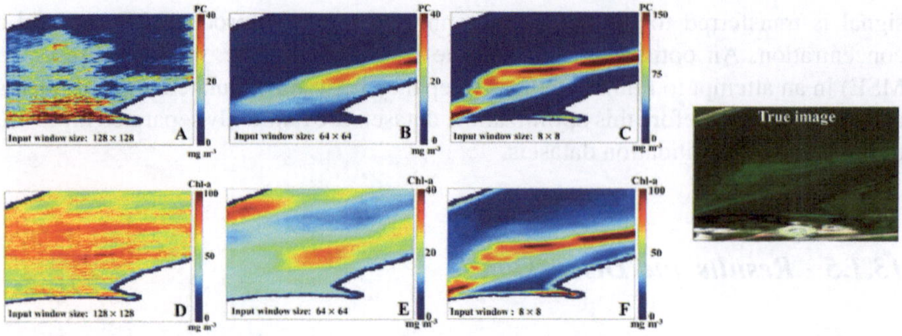

Fig. 13.4 PC and Chl-a map with different input window sizes, A and D are PC and Chl-a maps from 128 × 128 input window, B and E are PC and Chl-a maps from 64 × 64, and C and F are PC and Chl-a maps from 8 × 8

The influence of different filter sizes on the quantification of PC and Chl-a is illustrated in Fig. 13.4; the region with an algal bloom in the figure shows how the retrieval of PC and Chl-a are affected by the filter size. The model with the size of 128 × 128 showed the worst quantification of PC and Chl-a in comparison to the RGB map (Fig. 13.4A–B). The pigment maps with 64 × 64 resulted in better quantification but were inaccurate to reproduce the bloom areas (Fig. 13.4 C–D). It also underestimated PC and Chl-a in the area of maximum concentration nearby the gate. The model with 8 × 8 much improved the representation of PC and Chl-a showing similar spatial patterns with RGB maps (Fig. 13.4 E–F). This demonstrates that selecting the optimal size of the filter is a crucial step to extract the feature from images and to quantify algal pigments. It indicated that the smaller size resulted in the best representation, but it was not always the case. Sahiner et al. (1996) mentioned that large size in CNN would be more desirable for the extraction of deep feature from the objective of large-scale. It implies that there should be a necessary step to determine the optimal size with consideration of the size of corresponding or desired features from images. The performance of training and validation steps of PRCNN with 8 × 8 is shown in Fig. 13.5. Overall, the model showed satisfactory results for PC quantification, showing R^2 values of 0.90 (training) and 0.86 (validation) and RMSE values of 7.74 mg·m^{-3} (training) and 9.39 mg·m^{-3} (validation), even though PC estimated from the validation step slightly overestimated those from the training step.

The performance of Chl-a estimation is slightly poorer than the PC estimation, but it still shows acceptable accuracies in both steps (0.75 and 0.73 of R^2 values and 8.40 mg·m^{-3} and 8.34 mg·m^{-3} of RMSE values). The influence of the filter size explored in terms of validation error is shown in Fig. 13.6. It was observed that a relatively larger error was driven by the larger size of the filter (i.e., 64 × 64 and 128 × 128). Also, 8 × 8 was selected as the optimal size due to the small validation

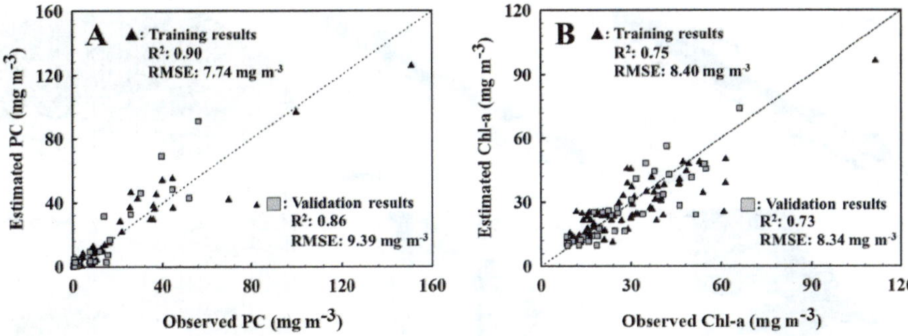

Fig. 13.5 Training and validation results of PC and Chl-a: A is the training and validation of PC, and B is training and validation of Chl-a

Fig. 13.6 Validation error with respect to different input window sizes as 4 × 4, 8 × 8, 16 × 16, 64 × 64, and 128 × 128

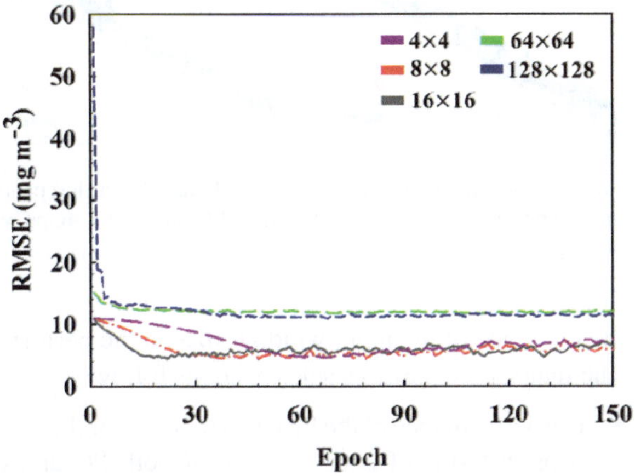

error. After training the model, the PRCNN model was used to generate the spatial distribution of PC and Chl-a concentrations for August 12, 2016 and August 24, 2016 (Fig. 13.7). The result showed that the maps generated from the model represent well the spatial distribution of algal pigments compared with RGB images.

13.1.6 Conclusion

This study applied the PRCNN with the hyperspectral images for quantifying PC and Chl-a in surface waters of Korea. The input included multiple images of atmospheric correction parameters and digital numbers, and the PRCNN model was trained and validated by the input information. After all processes, the PRCNN model finally was used to generate maps of PC and Chl-a pigments. In addition, this study explored

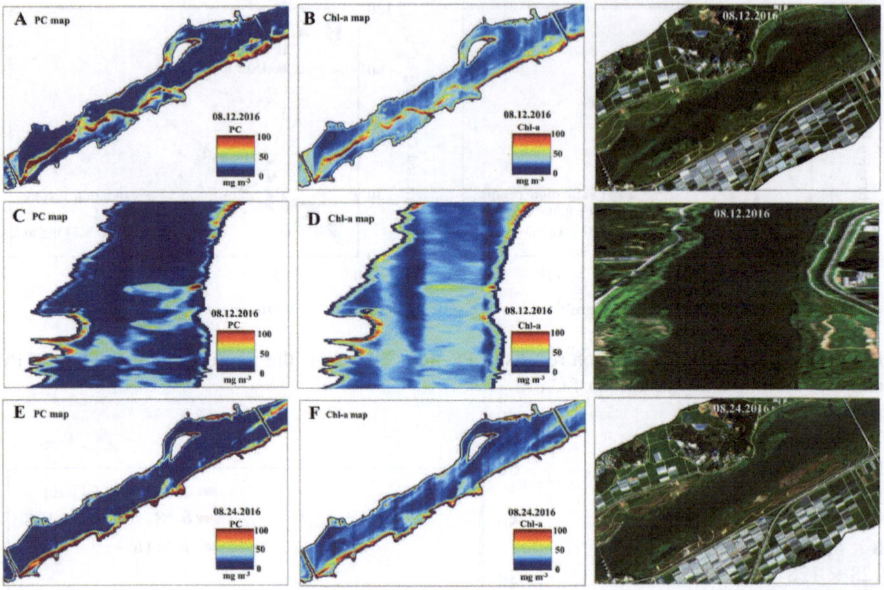

Fig. 13.7 Algal pigment map on August 12 and 24: the left panels (A, C, and E) are PC concentration maps, and the middle panels (B, D, and F) are Chl-a maps, while the right panels are true aerial photography images

the influence of the input window size on the performance of PRCNN performance. The major points of this study were as follows:

- The performance of the PRCNN model is highly accurate in training and validation steps in terms of concentration of both PC and Chl-a, implying that the deep-learning technique would be an alternative way to quantify the variation of PC and Chl-a.
- The PC and Chl-a map generated by the PRCNN model well reproduces the spatiotemporal variation of algal pigments with the small size of the input image (i.e., 8×8)

13.1.7 Python Coding

Below is the python code to import libraries related to TensorFlow, operating system, scipy, and numpy, and to determine the training parameters, including learning rate, training epochs, and batch size. After that, the code addressed two different directories; one is for saving the CNN model and the other is the location of existing data.

```
Python 13.1] CNN model python code
# Import library
import tensorflow as tf # Importing TensorFlow library
import os # Importing operating system library
import scipy.io # Importing python library used for scientific computing and technical computing.
import numpy as np # Importing fundamental package for scientific computing.

# Training parameters
learning_rate = 0.0001 # Setting learning rate
training_epochs = 150 # Setting epoch size
batch_size = 16 # Setting batch size of input

# Directory setting
model_path = "t7small8/model.ckpt" # Setting directory for trained model saving
DATA_PATH1 = os.path.join(os.getcwd(),"wq_1130_plz") # Setting current directory
```

The following code imports HSI, PC and Chl-a data from the directory assigned by the previous code where all data, imported from the directory, were separated into training and validation data sets. After that input and output spaces were created by considering their own specific dimension (i.e., 8 × 8 × 256 of input data). Additionally, the dropout probability was also determined.

```
[Python 13.1] Continue...
# Data importing
parm_total_tr = scipy.io.loadmat(os.path.join(DATA_PATH1, 'parm_sm8_tr.mat'))['parm_total_tr'] #
Import input imagery for training
wq_tr_pc = scipy.io.loadmat(os.path.join(DATA_PATH1, 'pc_label_tot_tr.mat'))['pc_label_tot_tr'] # Import
observed Phycocyanin concentration for training
wq_tr_chla = scipy.io.loadmat(os.path.join(DATA_PATH1, 'chla_label_tot_tr.mat'))['chla_label_tot_tr'] #
Import observed Chlorophyll-a concentration for training

parm_total_val = scipy.io.loadmat(os.path.join(DATA_PATH1, 'parm_sm8_val.mat'))['parm_total_val'] #
Import input imagery for validation
wq_val_pc = scipy.io.loadmat(os.path.join(DATA_PATH1, 'pc_label_tot_val.mat'))['pc_label_tot_val'] #
Import observed Phycocyanin concentration for validation
wq_val_chla = scipy.io.loadmat(os.path.join(DATA_PATH1, 'chla_label_tot_val.mat'))['chla_label_tot_val']
# Import observed Chlorophyll-a concentration for validation

# Making input and output data space
W1 = tf.placeholder("float", [None, 8, 8, 516]) # Space for the image data with 8 by 8 pixel window
Z1 = tf.placeholder("float", [None,1]) # Space for Phycocyanin data
Z2 = tf.placeholder("float", [None,1]) # Space for Chlorophyll-a data
keep_prob = tf.placeholder("float") # Space for the dropout probability
```

Below is the python code to build two convolutional layers by assigning filter size, stride, padding, etc., followed by the batch normalization layers, as shown in Fig. 13.3, a max-polling layer was assigned with a dropout layer right after the second batch normalization layer. "Flatten function" was used to vectorize outputs in an attempt to achieve the same dimension with PC and Chl-a observations.

```
[Python 13.1] Continue...
# Building CNN model
# Conv # 1
A1 = tf.layers.conv2d(inputs=W1, filters=32, kernel_size=[2, 2], strides=(1, 1), padding='valid',
                activation=None, name="Aconv1") # Building convolutional layer #1
A1_B = tf.contrib.layers.batch_norm(inputs=A1, decay=0.9, center=True, scale=False, epsilon=0.001,
                activation_fn=tf.nn.crelu) # Building batch normalization layer #1
# Conv # 2
A1_2 = tf.layers.conv2d(inputs=A1_B, filters=64, kernel_size=[2, 2], strides=(1, 1), padding='valid',
                activation=None, name="Aconv1-1") # Building convolutional layer #2
A1_2B = tf.contrib.layers.batch_norm(inputs=A1_2, decay=0.9, center=True, scale=False, epsilon=0.001,
                activation_fn=tf.nn.crelu) # Building batch normalization layer #2
A1_P = tf.layers.max_pooling2d(A1_2B, [2, 2], [2, 2]) # Building max-pooling layer
A1_P = tf.nn.dropout(A1_P, keep_prob=keep_prob, name="Adrop1") # Building dropout layer
fc = tf.contrib.layers.flatten(A1_P) # Vectorize the outputs (A1_P)
```

Below is the python code to generate two of sequential fully-connected layers (Fig. 13.3); one is for quantifying the PC concentration, while the other is for Chl-a concentration and each layer has the dropout option. The final products of the code will be PC and Chl-a concentrations (e.g., Y_pred1 and Y_pred2) calculated from the CNN model.

```
[Python 13.1] Continue...

dense1 = tf.layers.dense(inputs=fc, units=64, activation=tf.nn.crelu, name="Dense1") # Building fully
connected layer #1 for Phycocyanin
dense1 = tf.nn.dropout(dense1, keep_prob=keep_prob, name="Dense1_drop") # Building dropout layer

dense2 = tf.layers.dense(inputs=fc, units=64, activation=tf.nn.crelu, name="Dense2") # Fully connected
layer #1 for Chlorophyll-a
dense2 = tf.nn.dropout(dense2, keep_prob=keep_prob, name="Dense2_drop") # Building dropout layer

dense3 = tf.layers.dense(inputs=dense1, units=32, activation=tf.nn.crelu, name="Dense3") # Building fully
connected layer #2 for Phycocyanin
dense3 = tf.nn.dropout(dense3, keep_prob=keep_prob, name="Dense3_drop") # Building dropout layer

dense4 = tf.layers.dense(inputs=dense2, units=32, activation=tf.nn.crelu, name="Dense4") # Fully
connected layer #2 for Chlorophyll-a
dense4 = tf.nn.dropout(dense4, keep_prob=keep_prob, name="Dense4_drop") # Building dropout layer

Y_pred1 = tf.layers.dense(inputs=dense3, units=1, activation=None, name="Dense7") # Output for
estimated Phycocyanin concentration
Y_pred2 = tf.layers.dense(inputs=dense4, units=1, activation=None, name="Dense8") # Output for
estimated Chlorophyll-a concentration
```

Below is the python code to set up the loss function of the CNN model, to assign the optimizer, and to train the CNN model. The mean square errors of the Chl-a and PC estimation are set to be the loss function of the CNN model. The Adam optimizer was adopted to minimize the loss function in this code.

```
[Python 13.1] Continue...
# Loss function
loss = tf.reduce_mean(tf.square(Y_pred1 - Z1))+tf.reduce_mean(tf.square(Y_pred2 - Z2))  # Setting loss
function as a mean square error function for Phycocyanin and Chlorophyll-a

# Optimizer for model training
optimizer = tf.train.AdamOptimizer(learning_rate) # Adam optimizer adopted for the CNN model training
train = optimizer.minimize(loss) # The optimizer minimizes the loss function
train_size = parm_total_tr.shape[0] # Assign training size

# Proceeding CNN model training
with tf.Session() as sess:
   sess.run(tf.global_variables_initializer()) # Global variable initialization
   writer=tf.summary.FileWriter('./board/test5',sess.graph) # Operating Tensor board
   saver=tf.train.Saver(max_to_keep=150) # Operating model saver

   for epoch in range(training_epochs): # Initiating model training with respect to epoch
      avg_cost_1 = 0 # Initialize the cost for the loss function
      total_batch = int(train_size / batch_size) # Calculating batch size depending on the training data size

      # Training step in each iteration
      for i in range(total_batch):
         _, step_loss_1 = sess.run([train, loss], feed_dict={
            W1: parm_total_tr, Z1: wq_tr_pc, Z2: wq_tr_chla, keep_prob: 0.5}) # Proceeding training
                    with dropout rate as 0.5
         avg_cost_1 += step_loss_1 / total_batch # Calculating average errors in each batch size
error       in each epoch
      if epoch % 5 == 0:
         saver.save(sess, model_path, global_step=epoch) # Saving model in every 5 epoch
```

References

Chang JR, Chen YS (2015) Batch-normalized maxout network in network. In: Proceedings of the 33rd international conference on machine learning. W&CP, New York, JMLR

Chen Y, Jiang H, Li C, Jia X, Ghamisi P (2016) Deep feature extraction and classification of hyperspectral images based on convolutional neural networks. IEEE Trans Geosci Remote Sens 54(10):6232–6251. https://doi.org/10.1109/TGRS.2016.2584107

Clark JM et al (2017) Satellite monitoring of cyanobacterial harmful algal bloom frequency in recreational waters and drinking water sources. Ecol Ind 80:84–95. https://doi.org/10.1016/j.ecolind.2017.04.046

Duan H, Ma R, Hu C (2012) Evaluation of remote sensing algorithms for cyanobacterial pigment retrievals during spring bloom formation in several lakes of East China. Remote Sens Environ 126:126–135. https://doi.org/10.1016/j.rse.2012.08.011

Duan SB et al (2013) Land surface reflectance retrieval from hyperspectral data collected by an unmanned aerial vehicle over the Baotou test site. PLoS One 8(6). https://doi.org/10.1371/journal.pone.0066972

Glasgow HB, Burkholder JM, Reed RE, Lewitus AJ, Kleinman JE (2004) Real-time remote monitoring of water quality: a review of current applications, and advancements in sensor, telemetry, and computing technologies. J Exp Mar Biol Ecol 300(1–2):409–448. https://doi.org/10.1016/j.jembe.2004.02.022

Hu W, Huang Y, Wei L, Zhang F, Li H (2015) Deep convolutional neural networks for hyperspectral image classification. J Sensors https://doi.org/10.1155/2015/258619

Huang B, Collins LM, Bradbury K, Malof JM (2018) Deep convolutional segmentation of remote sensing imagery: a simple and efficient alternative to stitching output labels. Int Geoscience and Remote Sensing Symposium (IGARSS), pp 6899–6902. https://doi.org/10.1109/IGARSS.2018.8518701

Ioffe S, Szegedy C (2015) Batch normalization: accelerating deep network training by reducing internal covariate shift. Accelerating deep network training by reducing internal covariate shift, batch normalization

Kalchbrenner N, Grefenstette E, Blunsom P (2014) A convolutional neural network for modelling sentences. In: Proceedings of the conference on 52nd annual meeting of the association for computational linguistics. ACL, pp 655–665

Kellenberger B, Marcos D, Tuia D (2018) Detecting mammals in UAV images: Best practices to address a substantially imbalanced dataset with deep learning. Remote Sens Environ 216:139–153. https://doi.org/10.1016/j.rse.2018.06.028

Kimes DS, Knyazikhin Y, Privette JL, Abuelgasim AA, Gao F (2000) Inversion methods for physically-based models. Remote Sens Rev 18(2):381–439. https://doi.org/10.1080/02757250009532396

Kumar AC, Bhandarkar SM (2017) A deep learning paradigm for detection of harmful algal blooms. In: Proceedings on 2017 IEEE winter conference on applications of computer vision, WACV 2017, pp 743–751. https://doi.org/10.1109/WACV.2017.88

Lee DH et al (2014) Seasonal South Korea's trophic state index in the three weirs of Geum River. Korean Soc Anal Sci 2014:278

Lee Z, Carder KL, Chen RF, Peacock TG (2001) Properties of the water column and bottom derived from Airborne Visible Infrared Imaging Spectrometer (AVIRIS) data. J Geophys Res Oceans 106(C6):11639–11651

Mobley CD (1999) Estimation of the remote-sensing reflectance from above-surface measurements. Appl Optics 38(36):7442–7455

Pyo J et al (2019) A convolutional neural network regression for quantifying cyanobacteria using hyperspectral imagery. Remote Sens Environ 233. https://doi.org/10.1016/j.rse.2019.111350

Pyo JC et al (2018) High-spatial resolution monitoring of phycocyanin and chlorophyll-a using airborne hyperspectral imagery. Remote Sens 10(8). https://doi.org/10.3390/rs10081180

Sahiner B et al (1996) Classification of mass and normal breast tissue: A convolution neural network classifier with spatial domain and texture images. IEEE Trans Med Imaging 15(5):598–610. https://doi.org/10.1109/42.538937

Srivastava N, Hinton G, Krizhevsky A, Sutskever I, Salakhutdinov R (2014) Dropout: A simple way to prevent neural networks from overfitting. J Mach Learn Res 15:1929–1958

Yoon J et al (2015) Effect of hydrological conditions on distribution characteristics of algal concentration in the geum river main stream. Korean Soc Water Environ 2015:631–632

Zhang C et al (2018) An object-based convolutional neural network (OCNN) for urban land use classification. Remote Sens Environ 216:57–70. https://doi.org/10.1016/j.rse.2018.06.034